Gems of Hubble

Text by
Jacqueline Mitton
and
Stephen P. Maran

Foreword

The amazing discoveries made with the Hubble Space Telescope (HST) show how revolutionary advances in science occur whenever we greatly improve our technological capabilities or explore the universe in entirely new ways.

It was the power of the 60-inch and 100-inch telescopes at Mount Wilson, built early in the 20th century, that led astronomers to realize that our Sun is but one of billions of stars in a galaxy – a galaxy that is itself part of a universe containing billions of galaxies like our own. Radio telescopes, established in the 1950s, found that our Galaxy is filled with otherwise invisible clouds of cold hydrogen gas from which stars are born. They showed us the center of our own Galaxy for the first time, and revealed that the centers of many other galaxies continually release enormous amounts of energy. The combination of the new radio telescopes and what was then the largest optical telescope in the world, the Palomar 200-inch, resulted in the startling discovery of quasars in the 1960s. Radio telescopes also helped steer astronomers to one of the most fundamental discoveries of the 20th century, the observation that the universe is filled with ancient light released by the so-called 'Big Bang' from which the universe was born. Advances in technology have also increased our understanding of the solar system. Planetary probes and flybys launched in the last thirty years have revealed worlds previously unimagined. These examples all illustrate the power of new scientific tools to acquire new knowledge.

The idea of a large Earth-orbiting telescope was first proposed by Lyman Spitzer in 1946. 'The chief contribution of such a radically new and more powerful instrument' he wrote, 'would be, not to supplement our present ideas of the universe we live in, but rather to uncover new phenomena not yet imagined, and perhaps to modify profoundly our basic concepts of space and time'.

In 1977, when NASA asked astronomers to suggest scientific instruments for the HST, scientists immediately began to discuss the questions they hoped observations with the HST would answer. They agreed that one important question is, 'How fast is the universe expanding?' They also wanted to understand how galaxies evolve, and were curious about whether there are planets around other stars.

The HST was built during the 1980s and launched in April 1990. The years since have been momentous, beginning with the discovery of spherical aberration in the telescope's primary mirror, followed by the development of practical computer methods for improving or 'restoring' the out-of-focus images. Finally, in the successful HST First Servicing Mission in December 1993, teams of Space Shuttle astronauts installed corrective optics (COSTAR) for three of the instruments, replaced a camera with a new one (WFPC2) containing its own corrective optics, and made other necessary improvements. Today, the HST is performing nearly to perfection, as illustrated by the spectacular images in this book.

Most of the astronomers using the HST today were not born when Lyman Spitzer proposed the idea. With many years of useful life ahead of it, the HST will continue to enable men and women to explore the astronomical frontier into the 21st century.

—Robert A. Brown
Space Telescope Science Institute
April 1996

The HST Released in Space

The Hubble Space Telescope (HST) was launched from on-board the Space Shuttle *Discovery* on April 24, 1990. This photograph was taken by a *Discovery* astronaut shortly after the telescope had been released into a circular orbit 610 kilometers (380 miles) above the Earth. The solar arrays, which collect solar energy to power the telescope, have been unfurled. These are the paddle-like objects on either side of the telescope tube. The high-gain antennas, used to communicate with the ground, have also been extended. But the door over the end of the telescope tube is still closed to protect the delicate mirrors and instruments from contamination. The shiny door reflects clouds in Earth's atmosphere far below.

The HST was named in honor of the American astronomer Edwin P. Hubble (1889–1953), who discovered that the universe is expanding. Its main mirror is 2.4 meters (94 inches) across. When first launched in 1990 it was equipped with five instruments: two cameras for taking direct pictures, two spectrographs for analyzing light, and a high-speed photometer for measuring the brightness of stars and galaxies. The Fine Guidance Sensors, used for pointing the telescope very accurately, can also be used to measure the positions of celestial objects.

Credit: NASA

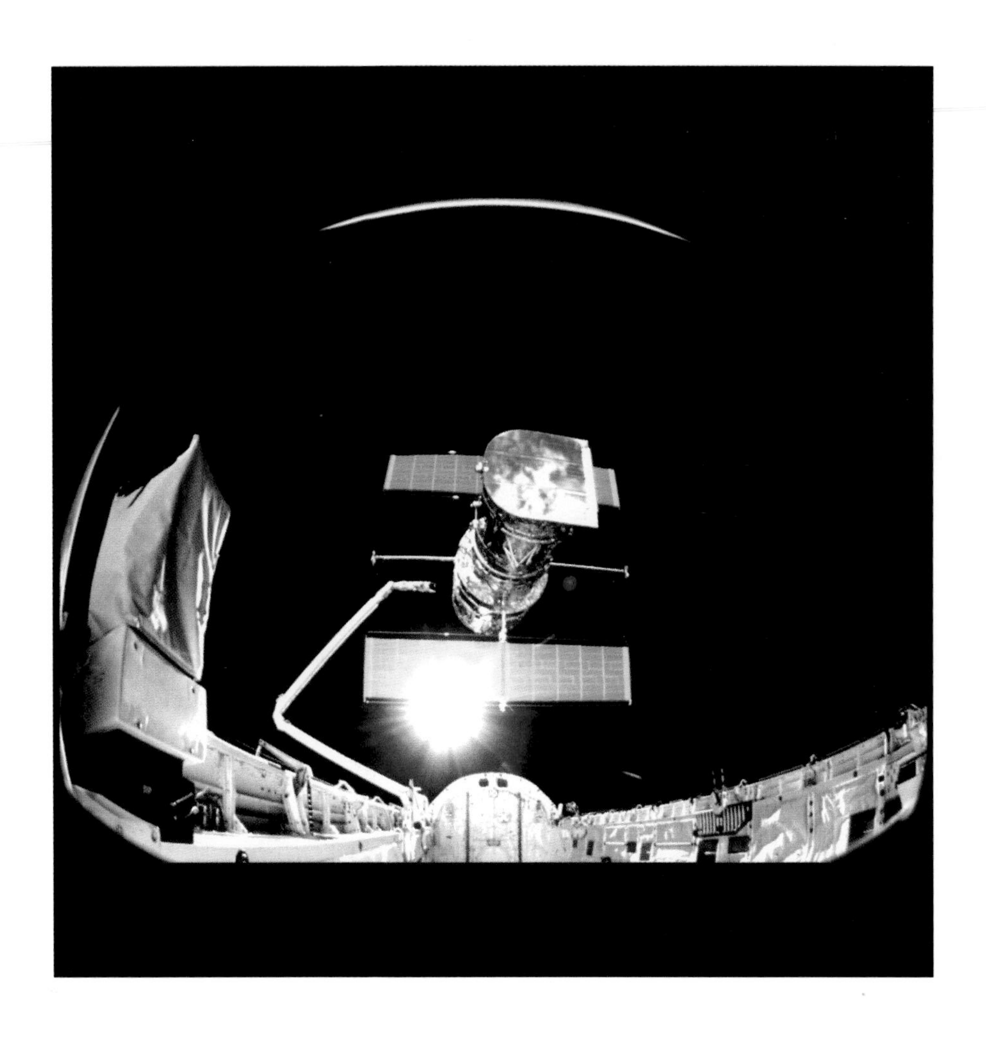

Replacing HST's Solar Arrays

HST's First Servicing Mission was carried out from the Space Shuttle *Endeavour* in December 1993. It involved five space walks. Here, during the second space walk, astronaut Kathryn Thornton prepares to replace the solar arrays. Her feet (not visible in the picture) are securely placed on *Endeavour*'s robot arm, which was used to carry the astronauts to each part of the HST in need of servicing. She is holding a hand-powered ratchet tool that she used to unbolt the arrays. Astronaut Thomas Akers participated in this space walk as well, but is not in the picture.

Behind Thornton, at the bottom of the image, is *Endeavour*'s open payload bay. Above and behind her is one of the old pairs of solar arrays that were replaced. It is partly wound back into its cylindrical container but would not retract completely because of a bent part.

The HST itself is further back in this picture, mounted on the floor of the payload bay, and pointing towards the Earth. On its special mounting, the telescope could be turned on command so the side the astronaut wanted to work on was facing him or her. The tube could also be tilted downwards. An electricity cable through the mounting supplied power to roll up the old solar arrays and roll out the new ones, and for other uses. *Endeavour*'s crew could switch as needed between this power supply and the HST's own six nickel–hydrogen batteries.

This image was taken during the daylight part of the orbit. The HST takes only about 90 minutes to circle the Earth during which time it goes through both daylight and darkness.

Credit: NASA

Removing HST's Old Camera

This picture was taken during the third space walk of the HST's First Servicing Mission. The original Wide Field and Planetary Camera (WF/PC-1) has been removed. Astronaut Jeffrey Hoffman, standing on the robot arm, grips it by the portable hand-hold, which has been bolted onto the camera during the space walk. A crucial piece of his equipment, the power ratchet tool, can be seen at his waist. He used it for unbolting the WF/PC-1 from the HST. Now he is preparing to clamp the camera temporarily on the grey bracket visible at lower left center. Later, WF/PC-1 was put into a box the size of a grand piano that had been used to bring the new camera, WFPC2, up from Earth. The old camera was brought back, first to the Goddard Space Flight Center, and then to the Jet Propulsion Laboratory, so it could be inspected and any reusable parts taken out.

The gold-colored object at the lower left is a camera, which can be operated both from the crew cabin on the Shuttle and from the ground. This camera supplied much of the video coverage of the space walks during the HST First Servicing Mission.

Astronaut Story Musgrave, who is not in the picture, was Hoffman's partner on this space walk.

Credit: NASA

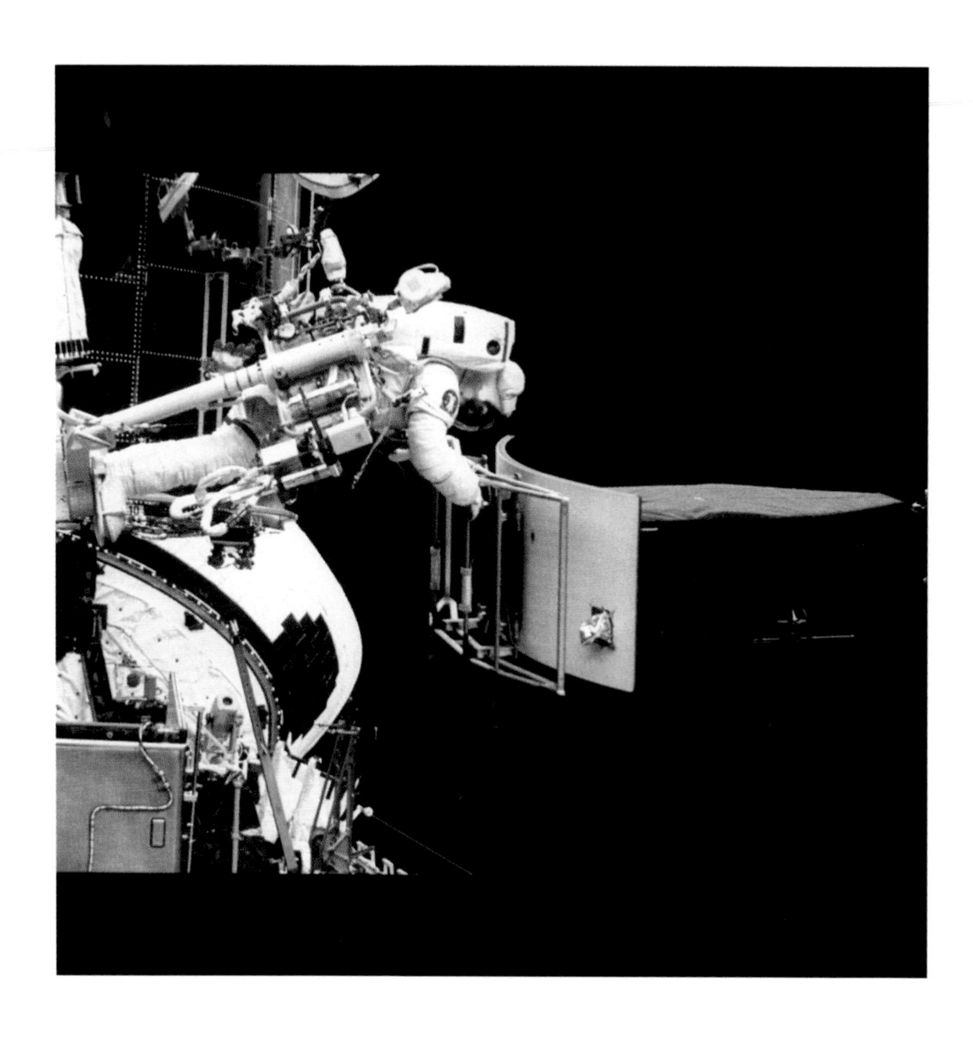

Star Images Before the HST's First Servicing Mission

This image was taken with the original Wide Field and Planetary Camera (WF/PC-1) before the First Servicing Mission. It shows stars located in the 30 Doradus nebula and star cluster, which is in the Large Magellanic Cloud. The Large Magellanic Cloud is a small galaxy neighboring our own Milky Way Galaxy at a distance of 169,000 light years. It is visible to the naked eye from the southern hemisphere.

The brightest star in the picture, known by the catalog number Melnick 34, appears surrounded by a distorted halo of light. This was due to the HST's focusing problem, technically called 'spherical aberration'. Light was spread over four arc seconds of sky instead of being concentrated into a sharp star image as it should have been. Fewer than 30 fainter stars can be discerned in this image.

To see how the view changed after WFPC2 was installed, turn to the next picture.

Camera: WF/PC-1
Credit: NASA

Star Images After the HST's First Servicing Mission

This image of stars in the 30 Doradus nebula and star cluster was taken with the new Wide Field and Planetary Camera, WFPC2, after the first HST Servicing Mission. It is the same area of sky covered by the previous picture. Small extra mirrors correct for the spherical aberration in the main mirror of the HST. Now the brightest star, Melnick 34, appears with its light accurately concentrated. The four spikes from the star image are a normal effect when a telescope like the HST images a bright star and are a sign of a well-adjusted optical system. Dozens of fainter stars are now revealed as well.

Camera: WFPC2
Credit: NASA

Springtime on Mars

It was spring in Mars's northern hemisphere when the HST took this view of the Red Planet on February 25, 1995. At the time, Mars was 103 million kilometers (65 million miles) away from Earth. The north polar ice cap visible here is several hundred kilometers across and made of solid water-ice. In winter, carbon dioxide frost settles and enlarges the polar cap but, at this time, the frost has gone.

On the western edge of the planet, where it is dawn, the surface is shrouded in clouds that have formed over night. When the temperature falls sharply at night, water in the atmosphere freezes to create clouds of ice crystals.

The summit of the extinct volcano, Ascraeus Mons, pokes through the cloud, appearing as a small oval shape near the left edge of the planet. It rises 25 kilometers (16 miles) above the surrounding plains and is 400 kilometers (250 miles) across. The dark feature running east–west towards the lower left is the Valles Marineris, a vast system of canyons extending for more than 5,000 kilometers (3,100 miles).

The dark areas were misinterpreted by early Mars watchers as regions of vegetation. In reality, they are areas of coarse sand that is less reflective than the finer orange dust. Seasonal changes in the appearance of the surface are the result of winds moving dust and sand around.

Camera: WFPC2 in PC mode.

Technical Information: Exposures through three color filters combined to create a true color image. Pictures map-projected onto a sphere for accurate registration and perspective.

Credit: P. James (University of Toledo), S. Lee (University of Colorado, Boulder), and NASA

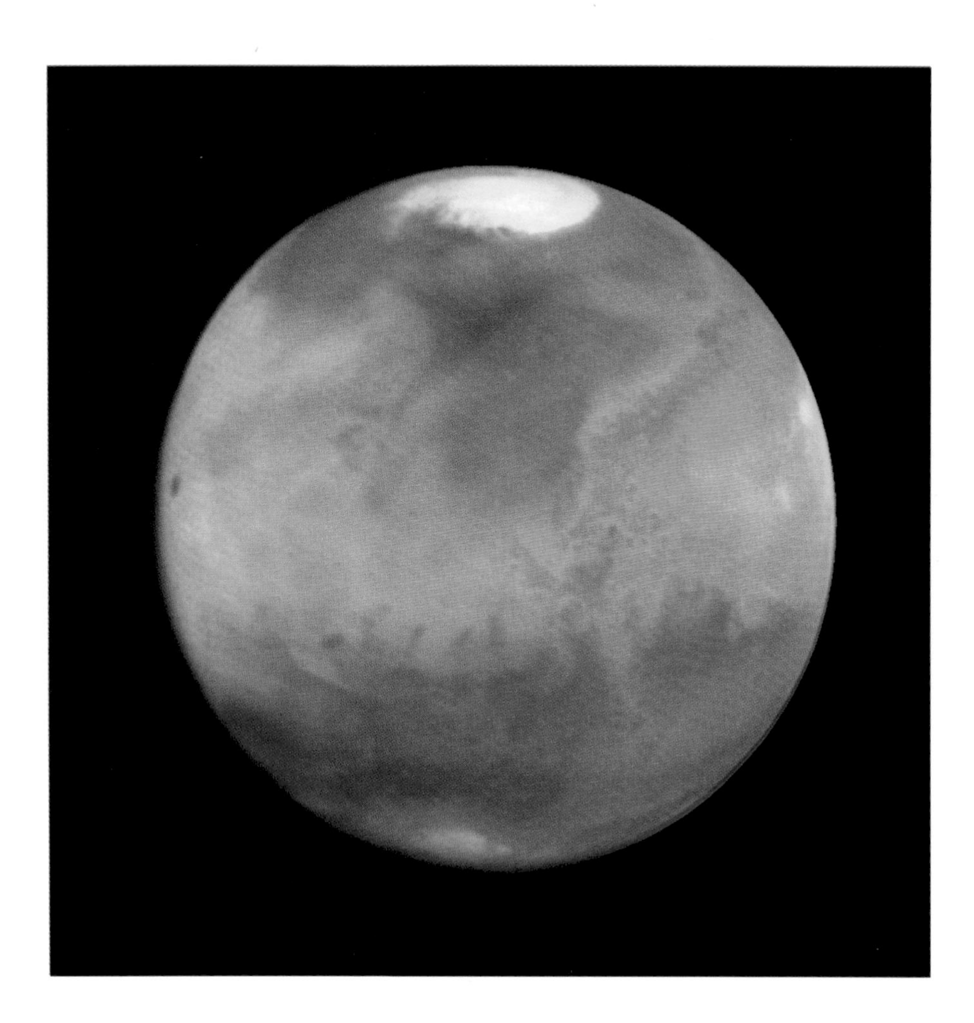

The Clouds of Venus

The surface of the planet Venus is permanently concealed from direct view by clouds of sulfuric acid in its thick carbon dioxide atmosphere. Ultraviolet images, like this one taken by the HST on January 24, 1995, reveal distinctive cloud bands that cannot easily be seen in visible light. In particular, there is a horizontal Y-shaped feature near the equator. Similar features have been observed by the spacecraft *Mariner 10*, *Pioneer Venus* and *Galileo*. Ultraviolet light is absorbed by Earth's atmosphere so images such as this can only be taken from space.

The brighter regions near the poles may be due to a haze of small particles above the main cloud layers. Darker areas show where sulfur dioxide is more concentrated near the cloud tops. Such features travel east-to-west, along with Venus's prevailing winds, making a complete circuit round the planet in four days.

Less than half of the disk of Venus is illuminated by sunlight in this image. Viewed from Earth – or near to the Earth as HST is – Venus goes through a cycle of phases similar to the Moon's because its orbit is nearer to the Sun than Earth's. As Venus gets closer to Earth during the course of its orbit, its disk appears larger overall, but the visible part of the illuminated side shrinks to a crescent. When this picture was taken, Venus was 113.6 million kilometers (70.6 million miles) away.

Though Venus's surface cannot be viewed directly, the *Magellan* spacecraft, operating in orbit around Venus between 1990 and 1994, used radar to map in detail the landscapes hidden beneath the clouds.

Camera: WFPC2 in PC mode.
Technical Information: False color has been used to enhance cloud features.
Credit: L. Esposito (University of Colorado, Boulder), and NASA

Jupiter

When the HST captured this image of Jupiter on May 18, 1994, its moon Io was crossing over the disk of the planet. The very dark circular spot on Jupiter is the shadow of Io. Had you been located on Jupiter's cloud tops inside the shadow, you would have experienced a total eclipse of the Sun. Io itself is the yellow-orange disk just to the upper right of the shadow.

Ten times the size of Earth, Jupiter is the largest planet in the solar system. It is made up almost entirely of hydrogen and helium. Clouds of water vapor, methane and ammonia float in the turbulent outer layers, creating the ever-changing pattern of swirling bands and spots. Jupiter's most famous cloud feature, the Great Red Spot, is just rotating into view on the left side of the disk.

This was the best image of Jupiter obtained since the spacecraft *Voyager 2* flew by the planet in 1979. The HST resolves details as small as 200 kilometers (124 miles) across even though Jupiter was 670 million kilometers (420 million miles) away.

Camera: WFPC2 in PC mode.

Technical Information: Exposures through three color filters (red, green and blue) combined to create a true color image. Jupiter's rotation between the exposures creates a blue and red fringe on either side of the disk.

Credit: H. A. Weaver and T. E. Smith (STScI), J. T. Trauger and R. W. Evans (JPL), and NASA

Io, Jupiter's Volcanic Moon

HST images of Io taken in 1994 and 1995, 16 months apart, reveal a change more dramatic than anything witnessed in the previous 15 years. A yellowish white circular feature 320 kilometers (200 miles) across developed where there was previously a much smaller spot.

Io is one of the four large moons of Jupiter discovered by Galileo in 1610. It is about the same size as Earth's Moon and circles Jupiter once every 42 hours and 28 minutes. In March 1979, the interplanetary space probe *Voyager 1* discovered large-scale volcanic activity on Io. Repeated eruptions and volcanic flows are continuously changing the surface, so its appearance varies noticeably from year to year. The vivid colored material and bright spots on Io are thought to be sulfur and chemical compounds of sulfur. The temperature on Io's surface is about –150°C (–238°F), but the hot-spots where there is volcanic activity may be as warm as 1,000°C (1,800°F).

The HST picture on the left, taken in March 1994, shows only subtle changes on Io since it was last seen in close-up, by *Voyager 2* in 1979. But by July 1995, the volcano Ra Patera was surrounded by new material from a large volcanic explosion or fresh lava flows, seen in the image on the right. The new bright spot is much yellower than other bright regions of Io, perhaps because the material is so fresh.

Camera: WFPC2 in PC mode.

Technical Information: Exposures through three color filters (near-ultraviolet, violet and yellow) combined to create color images.

Credit: J. Spencer (Lowell Observatory), and NASA

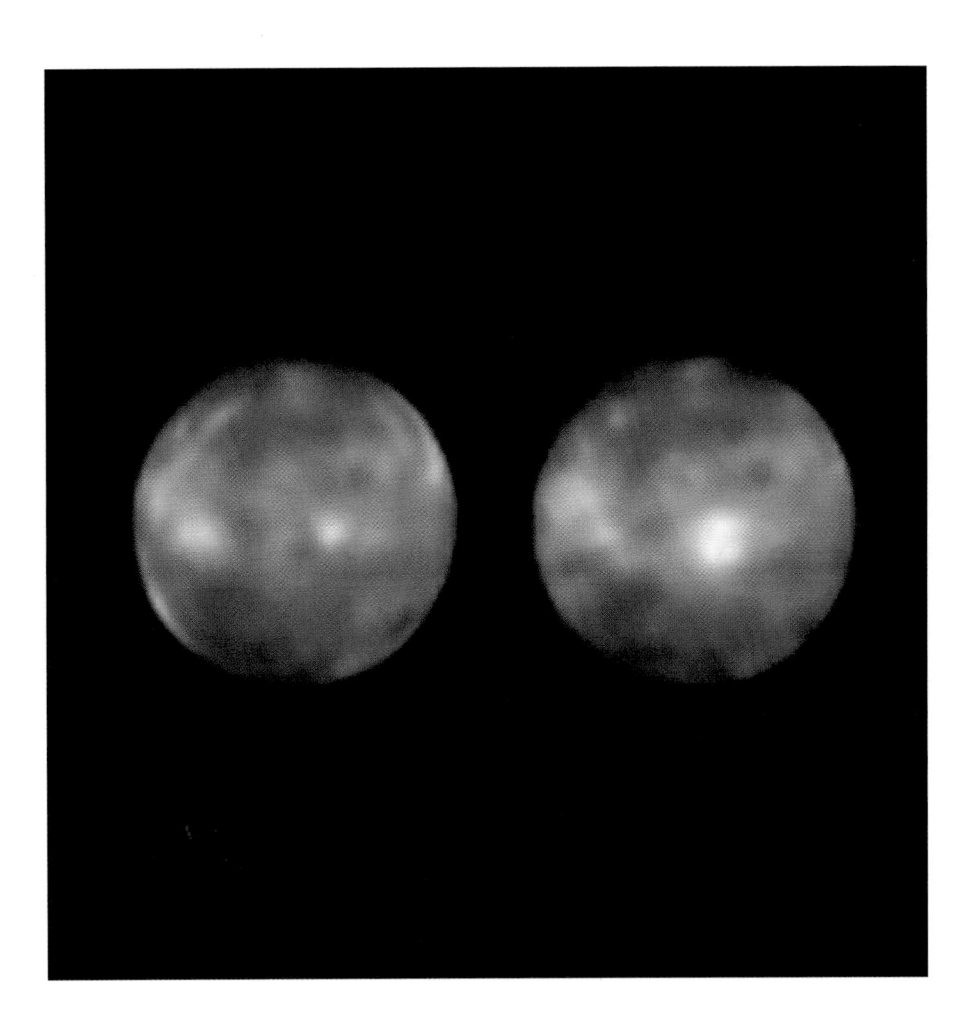

Saturn

A seemingly peaceful Saturn is depicted in this HST image of Saturn taken on August 16, 1990. The planet's north pole is tilted towards Earth at an angle of 24°, revealing atmospheric features in the 'northern polar hood'. Like Jupiter, Saturn is a gas giant, but there is much less detail and color contrast visible in its cloud bands.

Saturn's 'A' and 'B' rings are divided by the Cassini Division, the wide black band just outside the brightest part of the rings. The faint 'C' ring lies within the bright 'A' ring. The subtle Encke Division, a narrow dark gap near the outer edge of the outermost ring visible, is clearly seen. Although this thin gap was glimpsed through telescopes from Earth as early as 1837, it has not been successfully imaged from the ground.

Camera: WF/PC-1 in WF mode

Technical Information: Exposures through three color filters (blue, green and red) combined to create the color image.

Credit: NASA

Storm on Saturn

On rare occasions, the serenity of Saturn's atmosphere is disrupted by a significant storm. This image of one such storm was captured by the HST on December 1, 1994. It is the white feature shaped like an arrowhead lying close to the planet's equator. The extent of the storm, from east to west, was as great as Earth's diameter – about 12,700 kilometers (7,900 miles). It had changed very little since its discovery in September 1994.

Saturn's prevailing winds have shaped a dark 'wedge', seen here eating into the western (left side) of the bright cloud. The *Voyager* spacecraft measured the highest wind speed as 1,000 miles per hour at the latitude where the wedge has appeared. These strong winds blowing over the northern part of the storm created a disturbance that generated faint white clouds further east. To the north of the storm, where the winds decrease, cloud is being swept westward. The white clouds are made of ammonia crystals. They form when warmer gases from below push upwards through the cold cloud tops.

The HST observed a similar but larger storm on Saturn in September 1990. Very major storms such as the 1990 one seem to occur about once every 57 years, which is about twice the time it takes Saturn to orbit the Sun. White clouds associated with minor storms are reported more frequently, but the 1994 storm seen here is a relatively large and scarce event.

When this image was taken, Saturn was 1,446 million kilometers (904 million miles) from Earth.

Camera: WFPC2 in WF mode

Technical Information: Exposures through three color filters (blue, green and red) within 6 minutes combined to create a true color image. The blue fringe on the right limb is an artifact of the processing used to compensate for the rotation of the planet between exposures.

Credit: R. Beebe (New Mexico State University), D. Gilmore and L. Bergeron (STScI), and NASA

Dynamic Neptune

With the most detailed views seen since the *Voyager 2* spacecraft flew past Neptune in 1989, HST images have revealed how Neptune's appearance can change dramatically in only a few weeks. Bright clouds and dark spots appear and disappear regularly.

In June 1994, HST images showed that the Great Dark Spot seen by *Voyager 2* in Neptune's southern hemisphere had disappeared. By October–November, when these images were taken, a new dark spot had developed in the northern hemisphere, accompanied by bright high-altitude clouds made of frozen methane crystals. Gas flowing up over the spot cools as it rises and the crystals freeze out as the temperature falls.

Neptune is one of the gas giant planets with a diameter almost four times Earth's. Hydrogen and helium are its main constituents, but the visible atmosphere also contains methane and other gases. Methane, which absorbs red light and reflects blue light, contributes to Neptune's characteristic blue-green color. The pink features in these images are high-altitude clouds of methane ice crystals. Though they appear white in visible light, they are shown in pink here because they were imaged in the near infrared.

Earth's weather systems are driven by heat energy received from the Sun. The rapidly changing weather on Neptune, however, must be stirred up in a different way. Neptune has its own internal source of heat and radiates twice as much energy as it receives from the distant Sun. This means that Neptune's clouds are being warmed from below. Small variations in the temperature range between the top and bottom of the clouds could trigger the rapid large-scale changes witnessed by the HST.

Neptune was about 4.5 billion kilometers (2.8 billion miles) from Earth when these images were taken.

Camera: WFPC2

Technical Information: Picture reconstructed from images taken through filters at visible and near-infrared wavelengths.

Credit: H. Hammel (Massachusetts Institute of Technology), and NASA

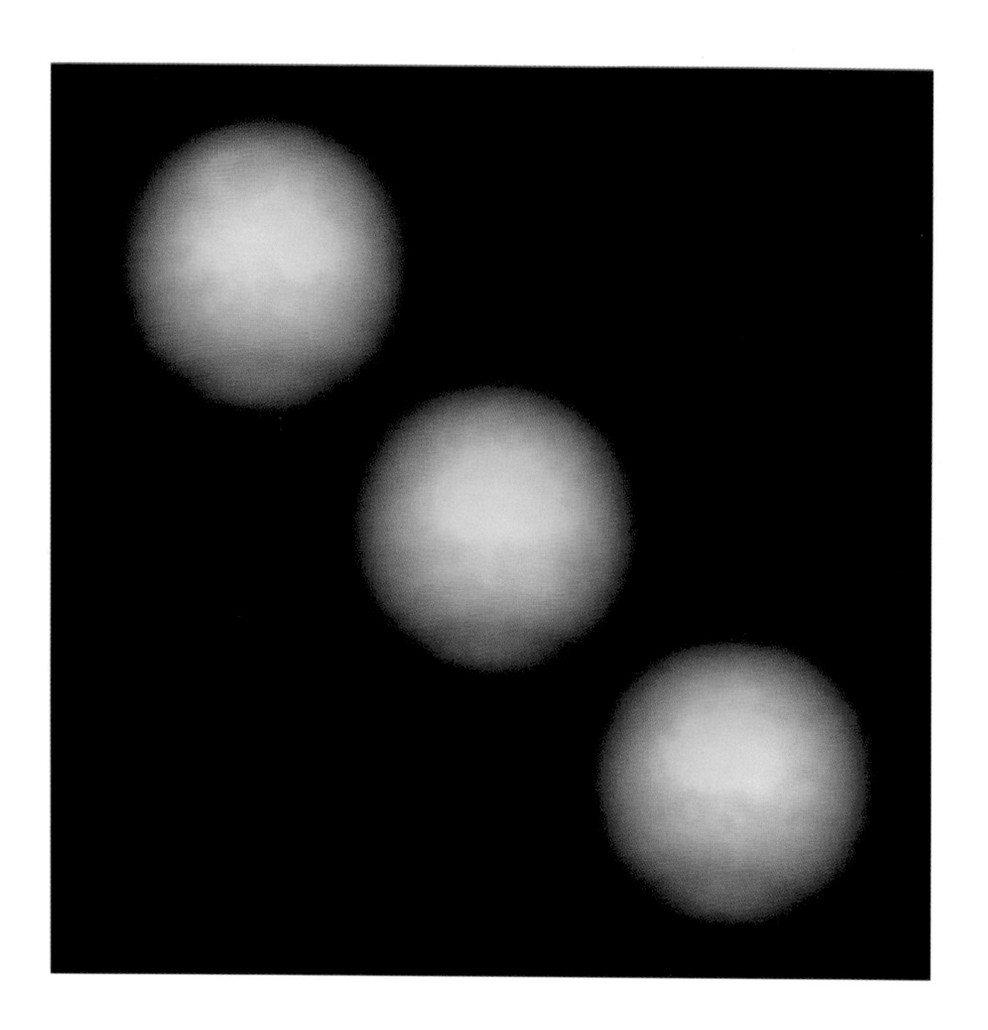

Pluto and Charon

Pluto – the smallest, coldest and most remote planet – made its closest approach to the Sun in 1989, about a year before the launch of the HST. With the help of the HST, astronomers hope to monitor changes in its thin atmosphere and icy surface as it moves outwards on its next 247-year orbit around the Sun. In this image, taken in July 1994 when Pluto was about 4,400 million kilometers (2,600 million miles) away, the large, northern polar ice cap was clearly revealed for the first time. Ground-based observations of the spectrum of Pluto indicate that the predominant icy material on the surface is frozen molecular nitrogen, N_2.

To show up the surface features more clearly, the contrast in this image has been exaggerated. The regions at mid and southern latitudes that appear dark are actually quite bright, reflecting 30–50% of the light falling upon them.

For 48 years after the discovery of Pluto in 1930, astronomers failed to see its moon, Charon, because the pair was very difficult to separate with the telescopes and techniques then available. It was not until 1978 that Charon was spotted by James Christy of the US Naval Observatory. Pluto is less than one fifth the size of Earth and Charon is just over one half the size of Pluto. The two are 19,640 kilometers (12,200 miles) apart. As shown here, the HST has no problem in revealing Pluto and Charon as distinct objects with a wide separation between them. However, no surface features can be made out on Charon.

Camera: Faint Object Camera with COSTAR

Technical information: Taken through blue filter (410 nm). Resolution is 12 to 15 resolution elements over the disk of the planet, almost 5 across a diameter.

Credit: S. A. Stern (SWRI), M. W. Buie (Lowell Observatory), L. M. Trafton (University of Texas at Austin), NASA, and ESA

Comet Shoemaker–Levy 9 a Year Before Impact

Comet Shoemaker–Levy 9 (SL9) was discovered close to the planet Jupiter on March 23, 1993, by Carolyn S. Shoemaker, Eugene M. Shoemaker and David H. Levy at Palomar Observatory and was named in their honor. This HST image of it was taken about three months later, on July 1, 1993.

It was a most unusual comet. At the time of its discovery, it was already in about 20 pieces, strung out across the sky like a row of pearls. Calculations soon showed that SL9 was in orbit around Jupiter, rather than around the Sun as comets normally are. What was more, it was on a collision course with the giant planet, due to crash through Jupiter's clouds in July 1994. In July 1992, SL9's orbit carried it so close to Jupiter that tidal forces tore it apart and sent the fragments on their course to final destruction. The break-up released large quantities of dust, which reflected sunlight well and made the comet easier to see. It is very likely that it had been orbiting Jupiter unnoticed for several decades before the fatal encounter of 1992.

Camera: WF/PC-1 in PC mode

Credit: H. A. Weaver and T. E. Smith (STScI), and NASA

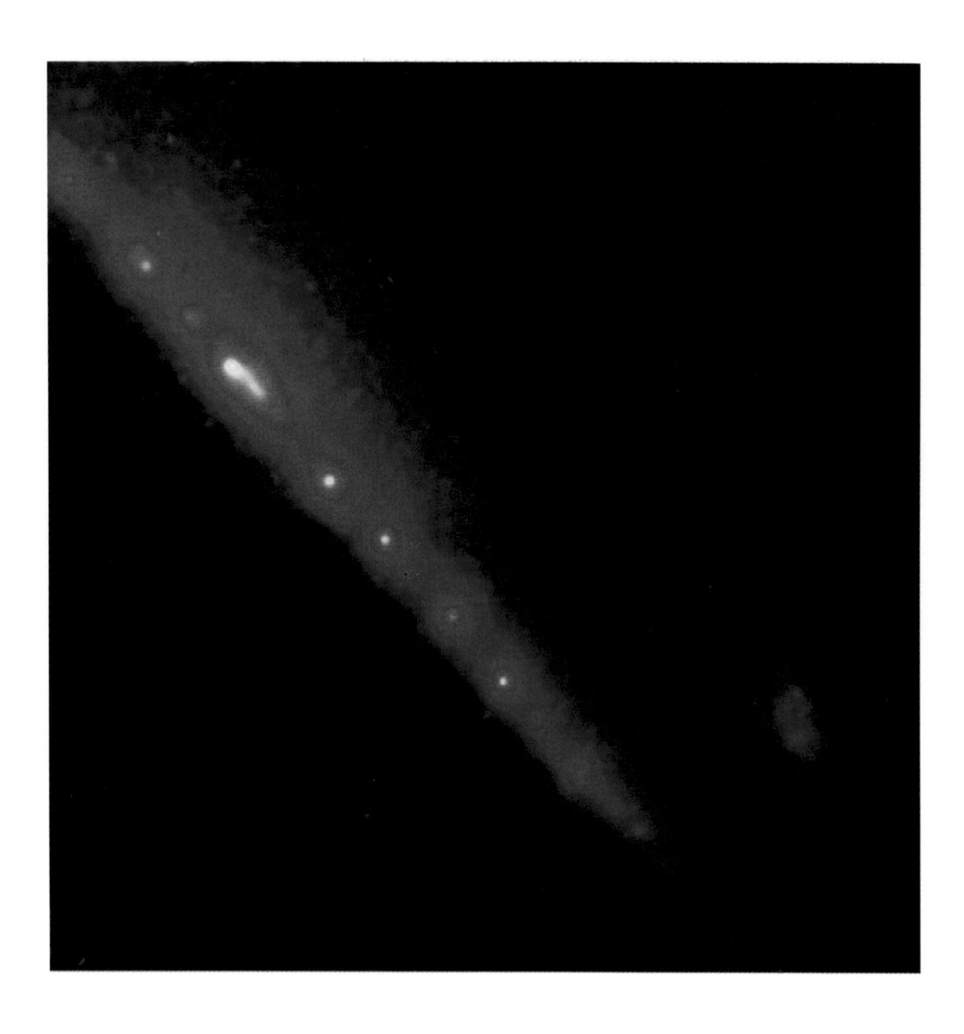

The Fragments of Comet Shoemaker–Levy 9

A mosaic of HST images (top) portrays almost the full extent of Comet Shoemaker–Levy 9 as it appeared in late January 1994. The train of fragments gradually lengthened, each separate piece following its individual orbit around Jupiter. Here it is almost 640,000 kilometers (400,000 miles) long. Each fragment is like a small comet with a nucleus, surrounded by a hazy coma, and its own tail.

Most of the fragments lie in a straight line, but several are clearly off the main train. These were probably formed when larger fragments from the original disintegration broke up again into smaller pieces, perhaps because they were spinning too fast to hold together, or because a pocket of gas erupted.

The three lower images show how one group of fragments broke up further and altered over a period of eight months. Formally named fragments P and Q, astronomers nicknamed them 'the gang of four' when it became clear that both were double.

In the image taken on July 1, 1993, before the optics of HST were corrected by the Servicing Mission (left), the double nature of the two bright nuclei is just detectable. With the new camera in operation, the image taken on January 24, 1994 (center) is sharper, and the two pairs of nuclei have also moved apart. By March 30, 1994 (right), the left-hand fragment of the lower pair has faded to a loose puff of dust. Meanwhile, the right-hand one has split into two, though only one of these survived to impact Jupiter in July 1994.

Camera: *Top:* WFPC2, mosaic of two images from WF components and one image from PC component.
Lower left: WF/PC-1.
Lower center and right: WFPC2.

Technical Information: Images shown in false color to distinguish different intensities of light.

Credit: H. A. Weaver and T. E. Smith (STScI), and NASA

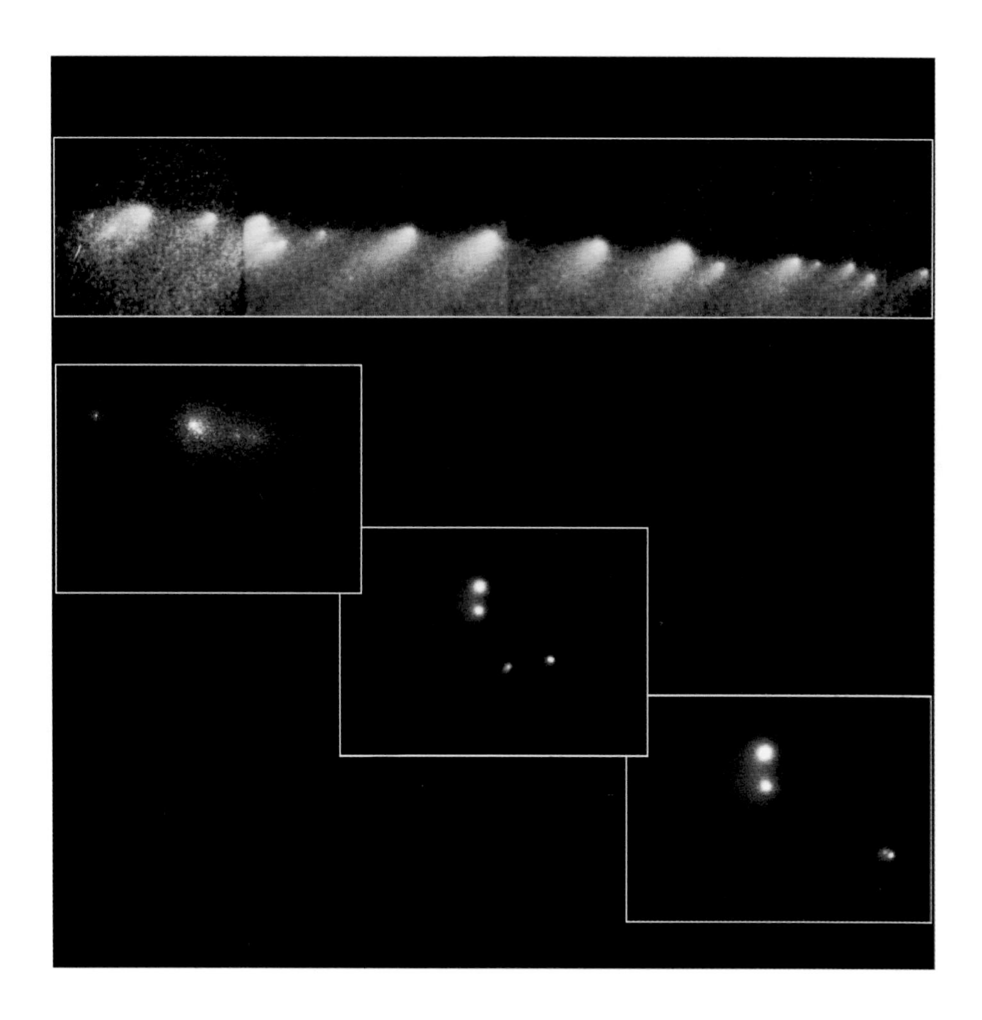

Impact Scars on Jupiter, July 1994

To the surprise of most astronomers, many of the fragments of Comet Shoemaker–Levy 9 created huge dark clouds of sooty particles in Jupiter's atmosphere when they crashed into the planet one at a time between July 16 and 22, 1994. At least eight impact sites are visible in this HST image of Jupiter, taken on July 21.

The fragments and their impact sites were identified by the letters of the alphabet – A, B, C, etc. In this image, the combined E and F sites are just visible on the left edge of the planet. The star-shaped H site is below and to the right of the Great Red Spot. Next comes a tiny dark spot marking the entry of fragment N, then the larger Q1 site, Q2 again small, R and at the far right, the complex created by impacts D and G.

The impacts all took place on the side of Jupiter facing away from Earth, but Jupiter's rotation carried the sites into view within a matter of minutes. All were at a jovian latitude of about 47°S but the sites were scattered all around the 47°S latitude circle because of Jupiter's rotation. Some of the fragments crashed into places where previous impacts had already created dark clouds, resulting in very complex scars. Soon, the high winds in Jupiter's atmosphere started to disperse the clouds. Six months after the impacts, very little evidence remained that they had taken place.

Camera: WFPC2

Technical Information: Exposures through three color filters (blue, green and red) combined to create a color image.

Credit: HST Comet Team, and NASA

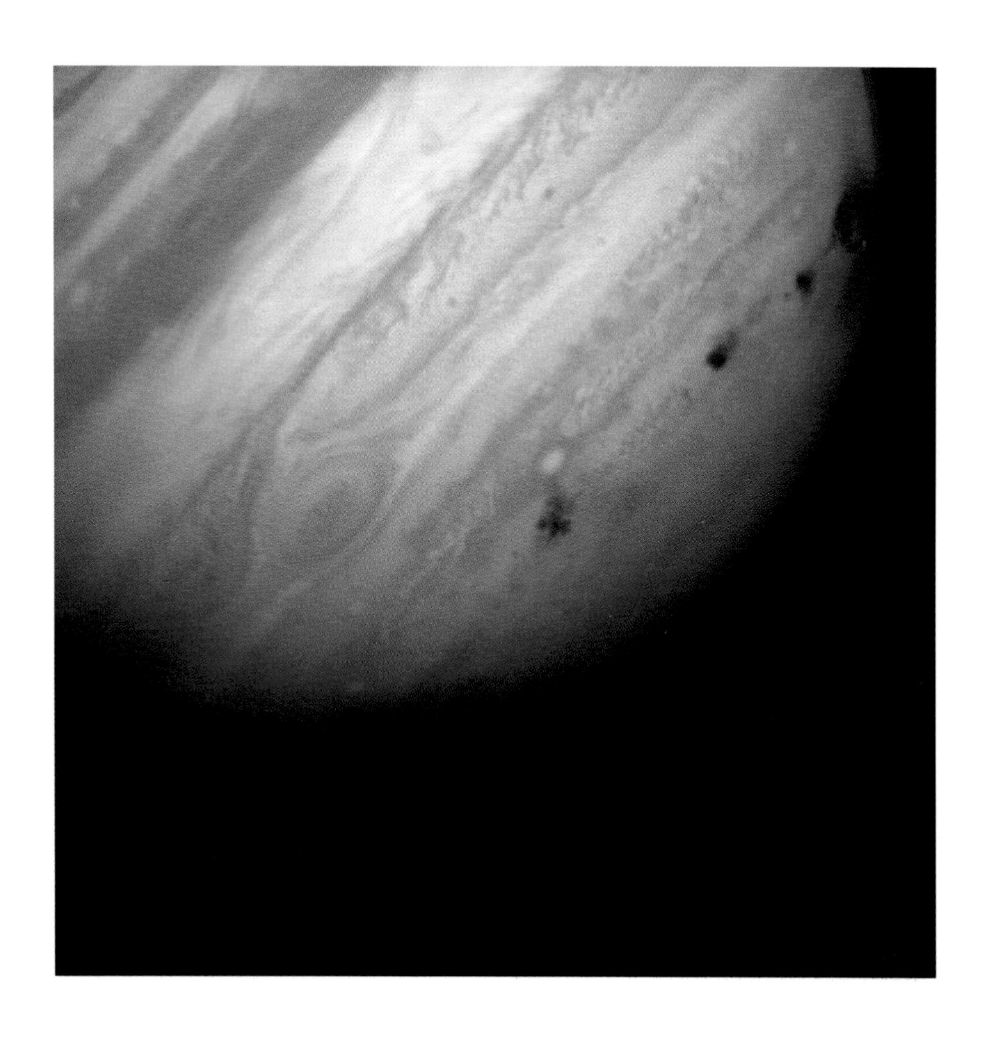

Asteroid Vesta

A series of 24 images taken by the HST between November 28 and December 1, 1994, are arranged here in order to illustrate one 5.34-hour rotation of the asteroid Vesta. Though only 525 kilometers (325 miles) across, Vesta is a true mini-planet which has survived almost intact since our solar system formed. This contrasts with many other asteroids which are chunks broken off larger objects when they were involved in collisions early in the history of the solar system.

Vesta's battered surface has been cratered by impacts and shows evidence of ancient lava flows dating from an era more than 4,000 million years ago when its interior was hot and molten. Since that time, Vesta has changed very little, apart from the effects of occasional meteorite impacts. The HST images reveal details down to 80 kilometers (50 miles) in size and provide a glimpse of some of the oldest terrain ever seen in the solar system. One large impact crater is so deep it seems to have torn away Vesta's outer crust completely, exposing the rocks of the mantle below.

Astronomers believe that fragments gouged out of Vesta during ancient collisions have fallen to Earth as meteorites, making Vesta only the fourth solar system object, after Earth, the Moon and Mars, from which scientists have a confirmed laboratory sample.

Vesta is the brightest known asteroid and the third largest. It was discovered in 1802 by H. W. M. Olbers and orbits the Sun in the main asteroid belt between Mars and Jupiter. When these images were taken, Vesta was 250 million kilometers (156 million miles) away from Earth.

Camera: WFPC2

Credit: B. Zellner (Georgia Southern University), and NASA

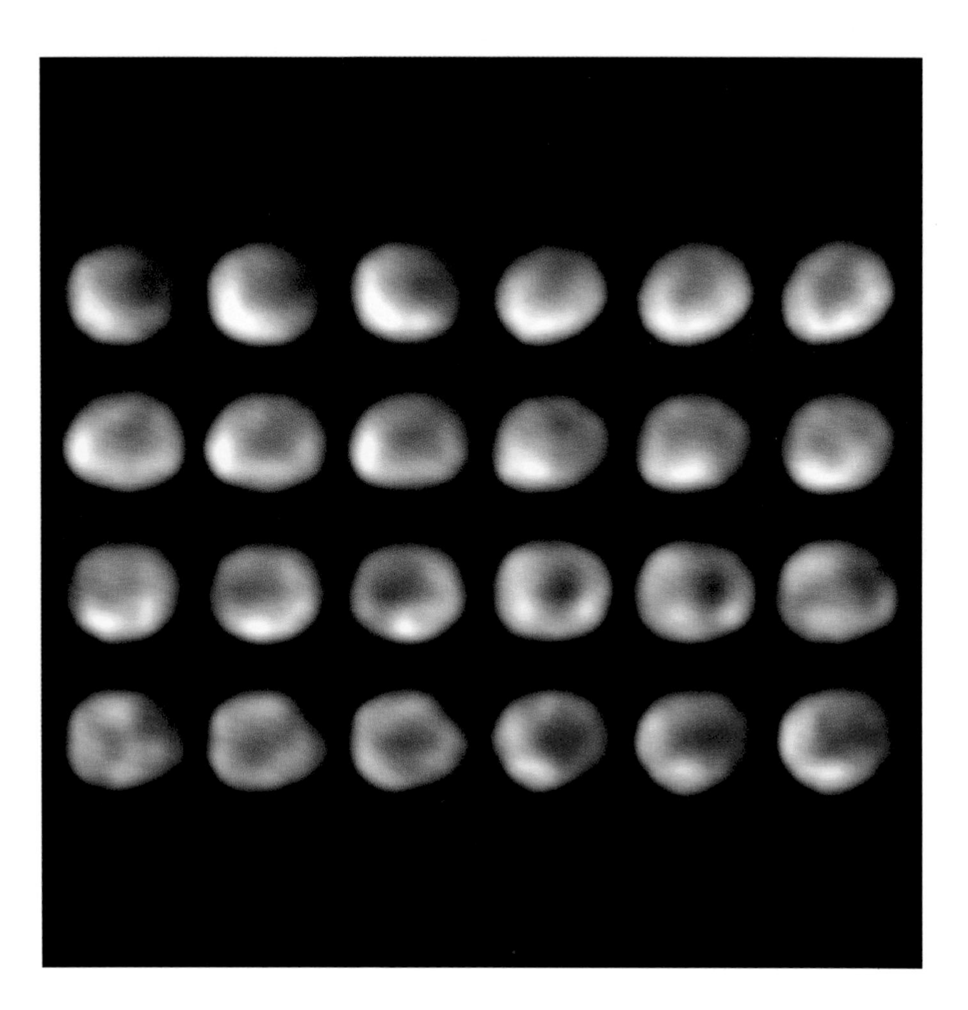

Newborn Stars Emerge in the Eagle Nebula

This dramatic picture reveals newborn stars emerging in the Eagle Nebula (also known as M16) in the constellation Serpens, a region of star formation 7,000 light years away. The newly forming stars are in oval-shaped clumps of interstellar gas appropriately named 'EGGs' – for 'evaporating gaseous globules'. Many of the EGGs are at the tips of finger-like features on great columns of cold gas and dust called 'elephant trunks', which protrude from the edges of a vast cloud of hydrogen molecules. Three 'elephant trunks' feature on this image. The tallest one, on the left, is about one light year long. Illuminated from behind by starlight, several EGGs can be seen at the top of this pillar of gas, looking as if they are just coming free.

Subject to a flood of ultraviolet light from young hot stars nearby (off the top edge of the picture), the surface of the molecular cloud is gradually being eroded in a process called 'photoevaporation'. The pillars seen here have survived longer than their surroundings because they are denser concentrations of gas. As the pillars themselves are eroded away, the small globules of even more dense gas growing inside them, the EGGs, are left behind. Eventually the EGGs break off. After separating, they cannot grow any more because they are isolated from the molecular cloud that was supplying them with new material. In time, even the EGGs succumb to photoevaporation and the newly forming stars inside them emerge.

EGGs at various different stages of development can be seen in the HST image. Some appear only as tiny bumps on the surfaces of the gas pillars. Others have been uncovered more and are joined to the pillar by a bridge of gas which has been protected from photoevaporation by the shadow of the EGG. Then there are some EGGs that have pinched off the pillar completely. In a few cases, the stars forming in the EGGs are directly visible.

Camera: WFPC2

Technical Information: Color image constructed from three separate images taken in the light of emission from singly ionized sulfur (red), hydrogen (green) and doubly ionized oxygen (blue).

Credit: J. Hester and P. Scowen (Arizona State University), and NASA

The Orion Nebula

The Orion Nebula is a glowing cloud of gas surrounding a cluster of newly formed stars about 1,500 light years away. It is just visible to the naked eye as a misty patch in the 'sword' region of the constellation Orion. This HST picture shows spectacular detail in the central part of the nebula. The main image covers a region about one and a half light years across.

The stars we see have formed from collapsing clouds of interstellar gas within the last million years. The most massive clouds have formed the brilliant stars near the center of the image, known as 'the Trapezium'. The ultraviolet light, mainly from the hottest and brightest star of all, illuminates the gas left behind after the period of star formation was complete.

The inset is a close-up of part of the main image, showing four young stars surrounded by clouds of gas and dust, believed to be protoplanetary disks, or 'proplyds'. These are flattened clouds of matter in which planets may form around the central star. Three of the proplyds in this image glow bright like the nebula itself. The fourth, on the right side of the picture, appears dark.

The HST images have revealed the presence of at least 153 proplyds in the Orion Nebula. This suggests that the formation of protoplanetary disks around stars is a common occurrence and is probably the rule rather than the exception. Seven of the proplyds are silhouetted against the bright background of the nebula. Estimates of their mass range from one tenth to 730 times the mass of Earth.

The proplyds visible in the main picture closest to the bright central star act like windvanes, with tails pointing away from the star. These tails are the result of light from the star pushing the dust and gas away from the outside layers of the proplyds.

This picture also shows the Orion Nebula to be a maelstrom of flowing gas in the form of curved shock waves, which are shaped by the same mechanism that forms shock waves around a supersonic aircraft in Earth's atmosphere.

Camera: WFPC2

Technical Information: Main picture (in original format) made from 45 HST images. Oxygen emission shown in blue, hydrogen emission in green and nitrogen emission in red.

Credit: C. R. O'Dell (Rice University), and NASA

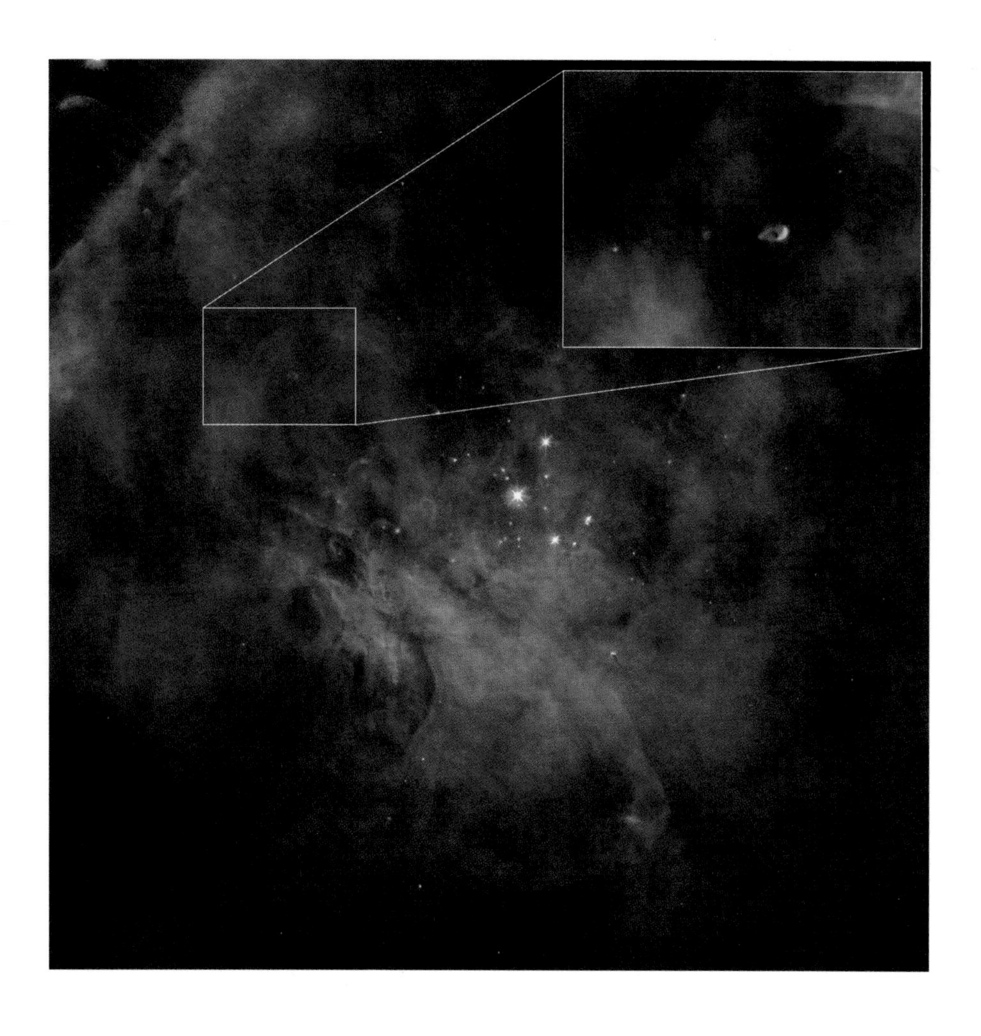

Jets from Young Stars

In the early 1950s, astronomers George Herbig and Guillermo Haro independently catalogued several enigmatic clumps of nebulosity close to stars near the Orion Nebula. These have since been called Herbig-Haro objects or 'HH' objects. Careful study later showed that many of the Herbig-Haro objects are portions of high-speed jets speeding away from newly-forming stars.

HH-47 in the top image is about half a light year long and 1,500 light years away. It lies at the edge of the Gum Nebula, possibly an ancient supernova remnant, in the southern constellations Vela and Puppis. The star responsible for the jet is hidden inside a dust cloud near the left edge of the image. The very complicated pattern in the jet suggests that the star might be wobbling, possibly because of the presence of a companion star. The jet has burrowed a cavity through the dense gas cloud. Shock waves form when it collides with interstellar gas causing the jet to glow. The white filaments on the left are reflecting light from the hidden star.

Underneath, the two images are HH-30 (left), which is 450 light years away in the constellation Taurus, and HH-34 (right), 1,500 light years distant in the vicinity of the Orion Nebula. The view of HH-30 reveals an edge-on disk of dust encircling a newly forming star. Light from the star illuminates the top and bottom of the disk while the star itself is hidden inside the densest part of the disk. The jet expands for several billion kilometers from the star, but then stays confined to a narrow beam. HH-34 shows a remarkable beaded structure, produced when blobs of dense gas are ejected from the star like bullets from a machine gun.

The bottom frame shows HH-1 and HH-2, the jets from a young star 1,500 light years away in the constellation Orion. The star is located midway between the blobs but is hidden from view behind a dark cloud of dust. Tip-to-tip they span slightly more than a light year. The pair of images above the bottom frame are close-ups of parts of it.

Camera: WFPC2

Credits: J. Morse/STScI, and NASA (HH-47); C. Burrows (STScI & ESA), the WFPC2 Investigation Definition Team, and NASA (HH-30); J. Hester (Arizona State University), the WFPC2 Investigation Definition Team, and NASA (HH-34, and HH-1/HH-2)

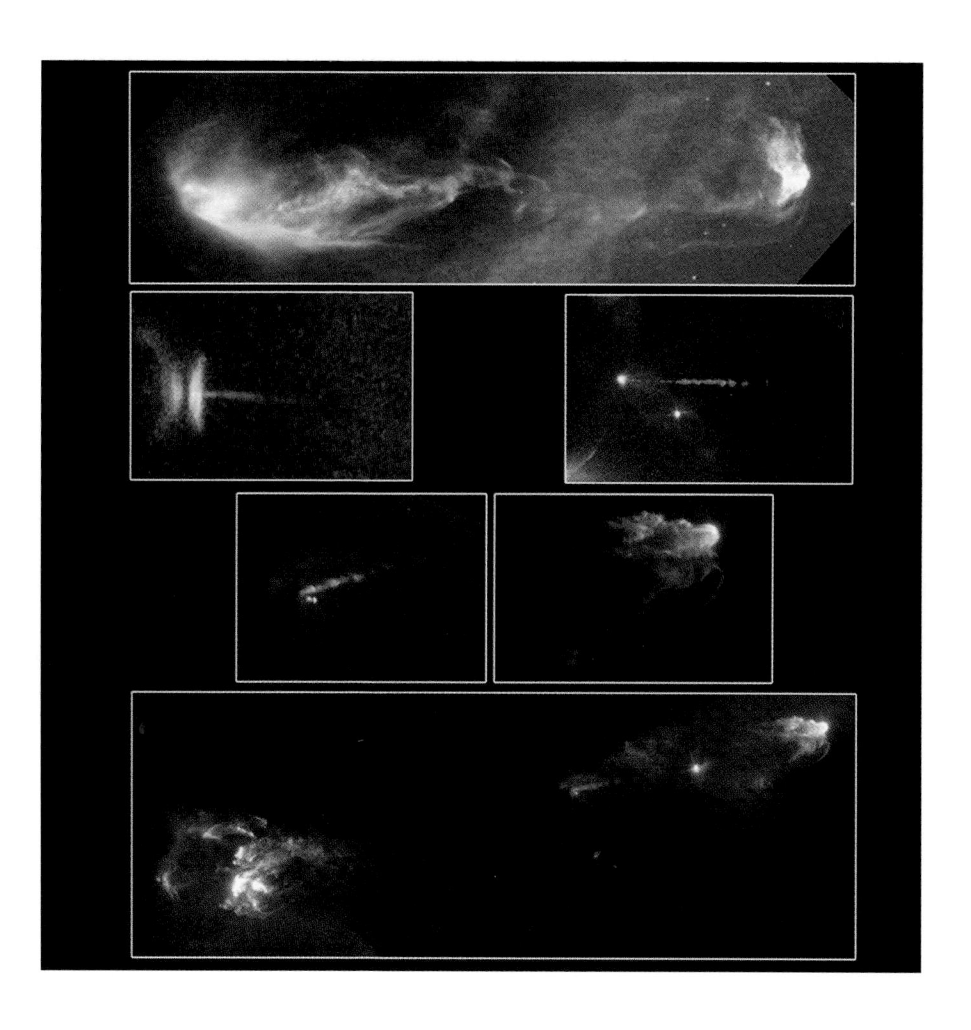

The Disk Around Beta Pictoris

This HST image revealed for the first time the inner regions of the disk of material that surrounds the star Beta Pictoris. From the ground, this inner part of the disk is impossible to observe because of the glare from the star itself.

The disk is made up of many microscopic grains of material, possibly a mixture of water ice and sand-like substances. This dust is lit up by the star. At the center of the disk is a clear zone and astronomers have speculated that this zone is occupied by a planetary system, similar to our own solar system. A planetary system could have formed from the dust in this part of the disk and would act to clear away any remaining material.

That idea is supported by evidence from this HST image. The inner edge of the disk is warped, in the same way that a CD would be warped if someone inserted a pencil in the central hole and twisted it. The lower version of the image has been further processed with false color to highlight the warp. Beta Pictoris is estimated to be 200 million years old, but a warp such as this could only last for about one million years before being flattened out – unless something is pulling on the disk and distorting it out of shape. A possible and plausible explanation is that a large planet, or a system of planets, is responsible for the warping action. It is not possible to see the suspected planet(s) directly because it (or they) are too close to the star, and perhaps a billion times fainter.

Attention was first focused on Beta Pictoris when it was found to emit infrared radiation strongly. Optical observations then showed there is a disk around the star with a diameter about ten times the size of Pluto's orbit around the Sun.

Camera: WFPC2 in PC mode

Technical Information: Composite of three images taken through filters at 555, 675 and 814 nm (green, red and infrared).

Credit: C. Burrows (STScI and ESA), J. Krist (STScI), and NASA

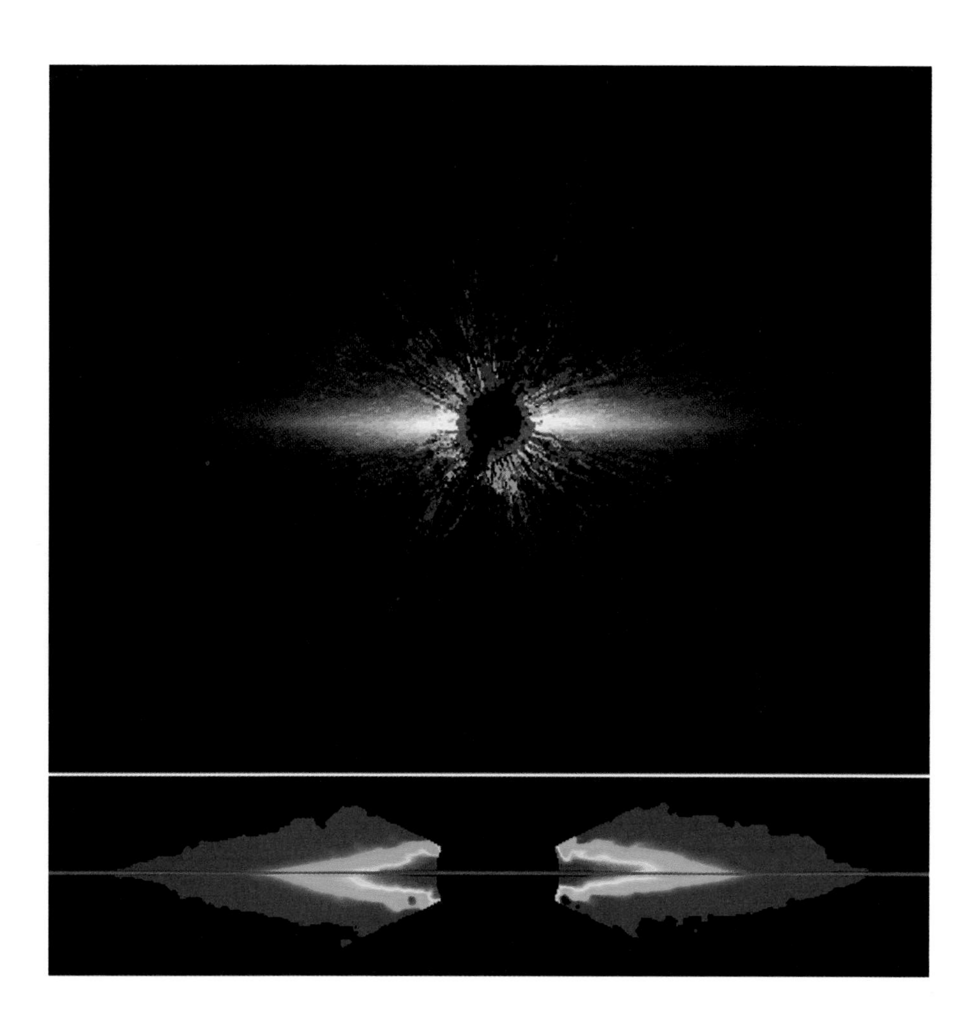

The Heart of the Globular Star Cluster 47 Tucanae

Many of the faint stars in this detailed ultraviolet image of the central part of the 47 Tucanae globular cluster are white dwarfs. These collapsed stars represent the final stage in the evolution of a star such as the Sun. Typically, a white dwarf is about the size of planet Earth and so dense that one cubic centimeter would weigh a ton. Its nuclear fuel has been exhausted so a white dwarf simply radiates its heat away into space, and gradually cools down over millions of years. Finding a group of white dwarfs in a cluster helps understand this process, since they are all the same age and started with the same chemical composition. White dwarfs are so dim it has not been possible to detect many of them in star clusters with ground-based telescopes.

A globular cluster is a ball-shaped concentration of hundreds of thousands – sometimes millions – of stars. 47 Tucanae, which is 15,000 light years away, is one of several hundred such clusters known in our Galaxy. It is the second-brightest globular cluster in the sky, just detectable to the naked eye. The globular clusters in our Galaxy contain some of its oldest stars.

Ultraviolet HST images of the center of 47 Tucanae also reveal the presence of stars that are unexpectedly hotter and bluer than the rest of the stars in the cluster. Astronomers nick-name them 'blue stragglers'. They seem to be more numerous near the center of the cluster where the stars are packed most closely together. No-one knows for sure what they are but a possible explanation is that they are pairs of stars that linked up as close doubles, or perhaps even merged, as a result of close encounters.

Camera: Faint Object Camera with COSTAR
Credit: R. Jedrzejewski (STScI), NASA, and ESA

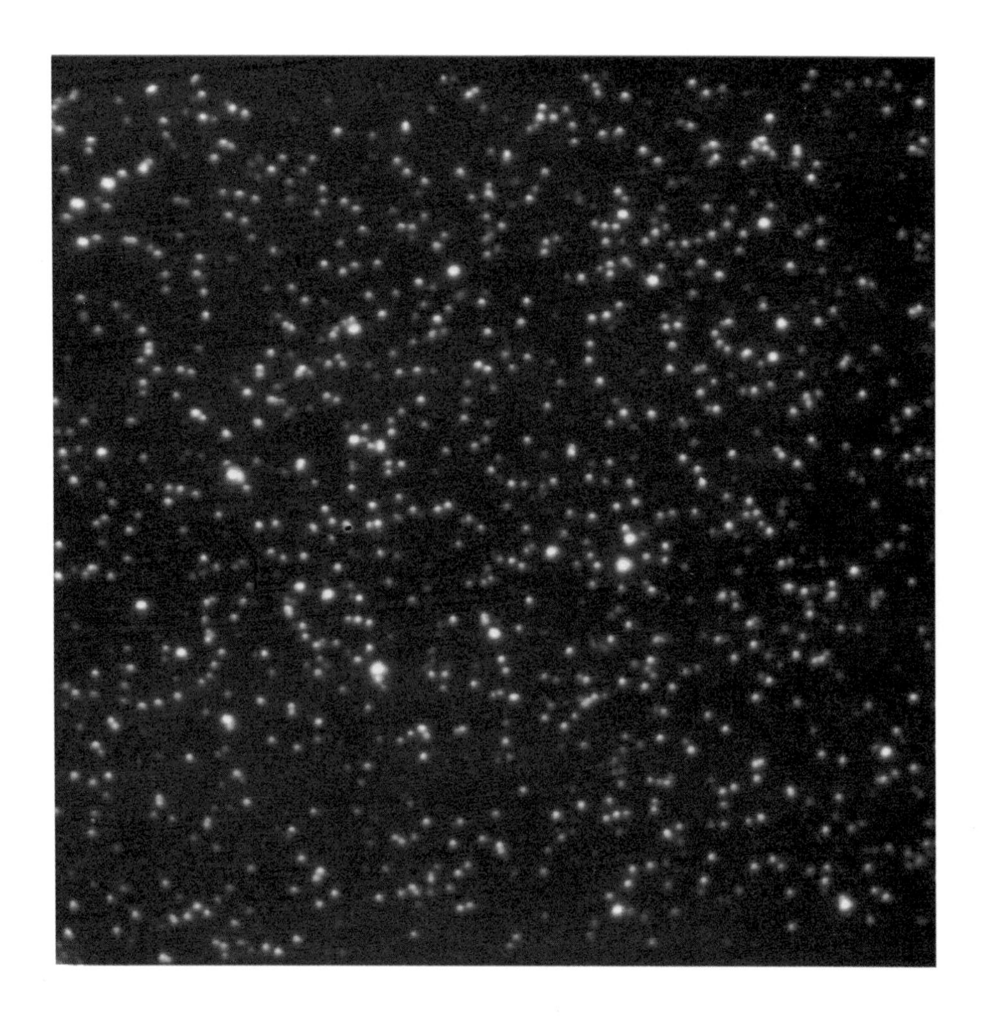

Two Star Clusters in the Large Magellanic Cloud

There are nearly 10,000 stars in this image of part of the Large Magellanic Cloud, spanning a region of space about 130 light years across. Over half of them belong to a cluster called NGC 1850, which is estimated to be 50 million years old. They appear predominantly yellowish. A scattering of white stars in the image are massive stars only about 4 million years old. Besides being much younger, the white stars are grouped much more loosely than the yellow ones. They account for a further twenty percent of the stars seen in this picture. The rest are so-called 'field stars' – stars that just happen to lie in the line of sight but do not belong to either of the clusters.

The significant age difference between the 'white' and 'yellow' clusters suggests that they are two completely separate clusters that just happen to lie along the same line of sight. The 'white' cluster probably lies about 200 light years beyond the other one. If it were in the foreground, dust within this young cluster would obscure the stars of the other cluster.

It is unusual to observe two such different populations of stars so close to each other in space. It suggests that a supernova explosion in the older cluster might have triggered the birth of the younger stars.

The Large Magellanic Cloud is a small galaxy neighboring the Milky Way at a distance of 169,000 light years. It is visible to the naked eye in the southern constellation Doradus.

Camera: WFPC2

Technical Information: Composite image assembled from exposures taken in ultraviolet, visible and near-infrared light.

Credit: R. Gilmozzi (STScI and ESA), S. Ewald (JPL), and NASA

Radcliffe 136: An Extraordinary Star Cluster

Radcliffe 136 (R136) is a compact cluster of hot young stars in the 30 Doradus nebula, an enormous region of glowing hydrogen gas in the Large Magellanic Cloud. At one time, the brightest object in R136 was believed to be a single superlative star with at least 3,000 times the mass of the Sun. However, according to astrophysical theory, a star cannot form with a mass of much more than 120 or 150 solar masses. A star of any greater mass would generate light at a rate so furious that the pressure of radiation flowing outwards would blow off any additional matter from the outside. So it was recognized that this supermassive object is in reality a tight cluster of individual stars. Prior to the first HST observations in 1990, the best observations, combined with mathematical theory, suggested there were at least 27 stars. Now HST has shown there are at least 3,000 stars in the vicinity.

Camera: WFPC2
Credit: NASA

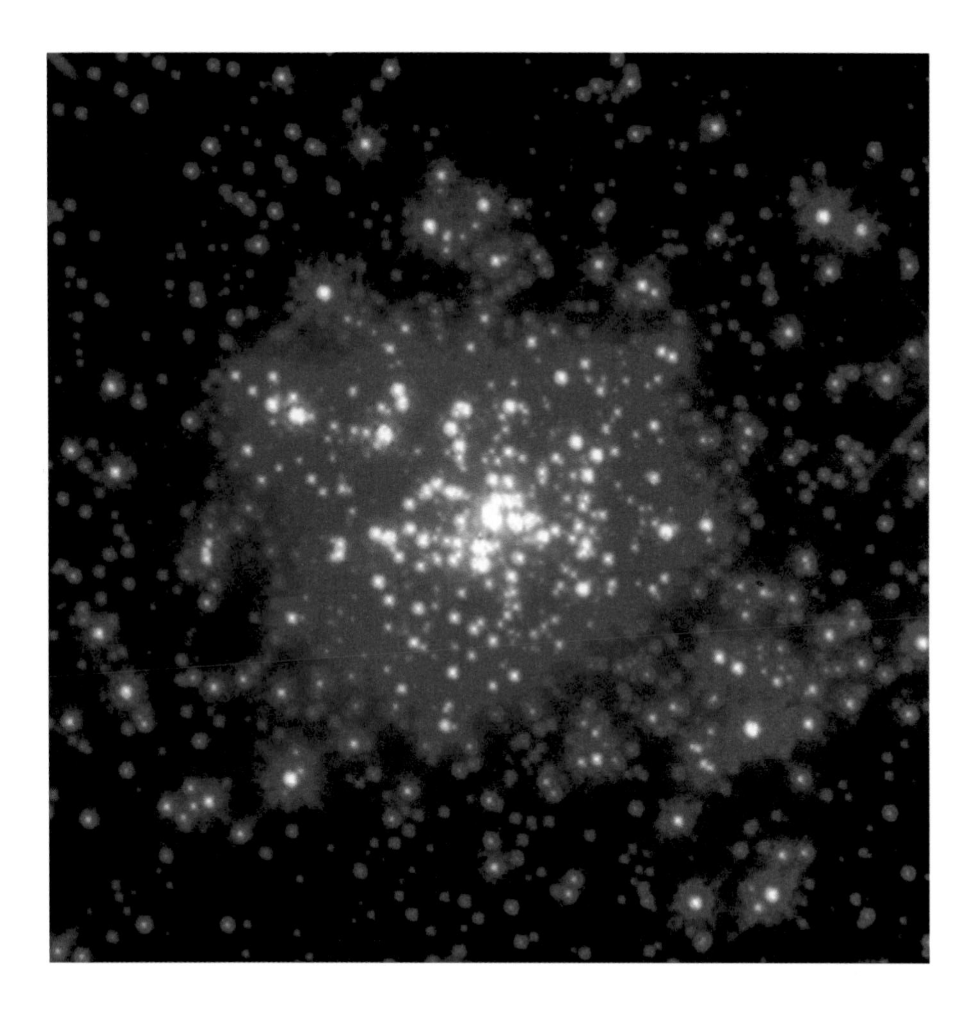

Star Cluster R136 in its Setting

The star cluster R136 is seen here at the upper right against the background of the spectacular 30 Doradus nebula (also known as the Tarantula Nebula) in the Large Magellanic Cloud. The 30 Doradus nebula is a cloud of glowing interstellar gas, energized by radiation from nearby hot, luminous stars. If the 30 Doradus nebula, which is 169,000 light years away, were put in the place of the Orion Nebula, at a distance of 1,500 light years, it would light up the night sky like a full Moon.

Camera: WFPC2
Credit: NASA

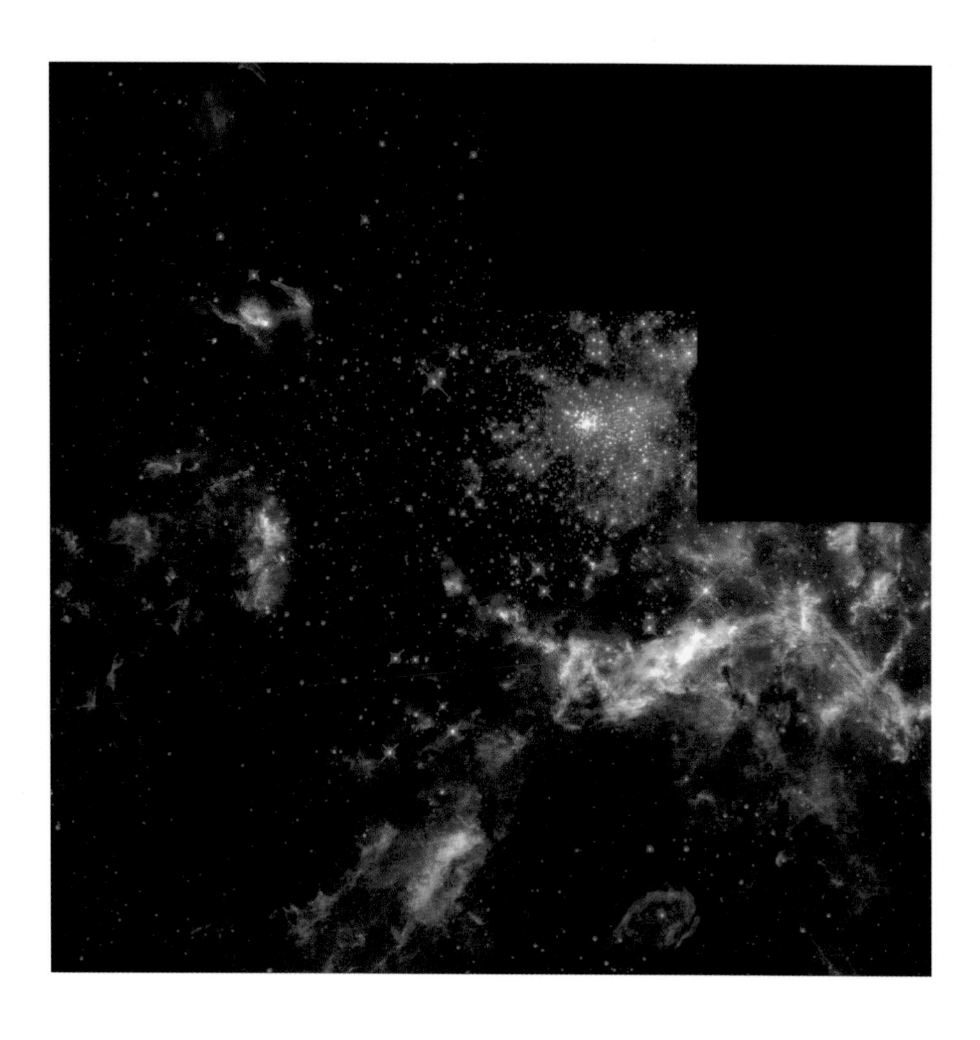

The Spectra of Two Stars in the Cluster R136

These spectra show ultraviolet starlight broken down into its component wavelengths or colors, in the same way that visible light is spread into its colors in a rainbow. With spectra like this, astronomers can investigate the chemical composition, temperature, density and motion of the stars.

The spectra of two stars in the cluster R136 are shown here, each in two different formats. First (top), the spectrum appears in 'rainbow' format. Bright vertical columns are particular wavelengths where emission is strong ('emission lines'). An absorption line, where certain atoms cut off or reduce the light at particular wavelengths, appears as a dark column. The second format presents the information as a graph with wavelength as the horizontal axis and the strength of radiation as the vertical axis. In this format, an emission line appears like a narrow hill and an absorption line appears as a dip or valley.

A strong emission line is visible in the spectra of both stars at a wavelength of 1645 Å (164.5 nm). It is identified as emission from singly ionized helium in the winds of gas blowing off the stars. These emission lines are telltale clues to the amount of matter blowing out of the stars. From this evidence, the star R136a5 appears to be blowing off a mass equal to the mass of the Sun in just 50,000 to 100,000 years. Such a rapid rate of mass loss greatly affects the future fate of the star.

Instrument: Goddard High Resolution Spectrograph (GHRS) with COSTAR
Credit: S. Heap/NASA

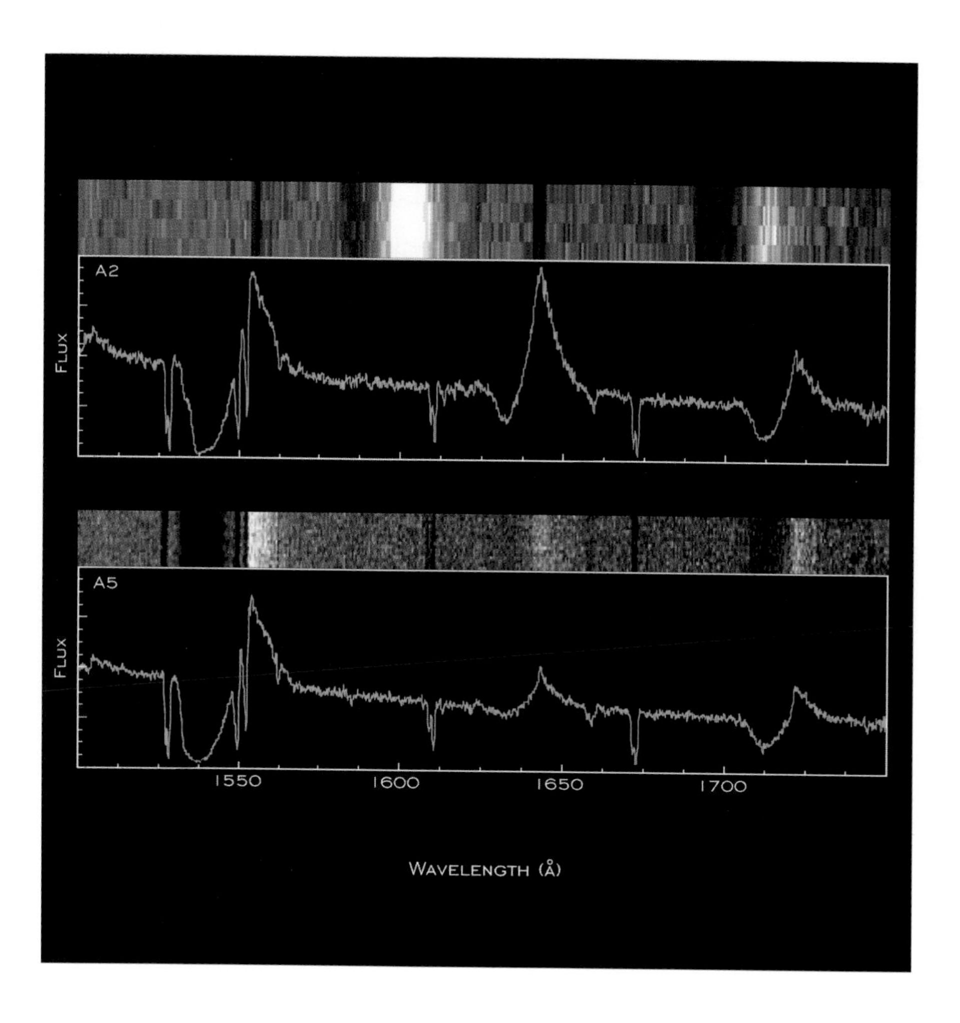
A2
FLUX
A5
FLUX
1550
1600
1650
1700
WAVELENGTH (Å)

The Fate of the Massive Star R136a5

This diagram shows two possible fates of a massive hot star such as R136a5. With a mass of about 60 times that of the Sun, R136a5 should exhaust its nuclear fuel and come to its end – whatever that might be – in just a few million years. The fate of this massive star may depend on the rate at which it is losing mass in its stellar wind.

If the rate of mass loss were low, R136a5 would expand and become a red supergiant star, as shown on the left side of the diagram. Soon, the star would use up its central storehouse of nuclear fuel and then collapse into a black hole. However, it appears that a slightly less exotic fate awaits the massive star, and there is no black hole in the offing. The HST observations of R136a5 revealed that it has a stellar wind so strong that gas equal in mass to our Sun is removed every 50,000 to 100,000 years. If the wind continues at that furious rate, the mass of R136a5 will be reduced so much by the time that its central nuclear fuel store has been used up, that it will explode as a typical 'Type II' supernova. The imploding stellar core will produce a neutron star.

In the interim, it appears that the rapid mass loss from the exterior of R136a5 may form an expanding cloud of gas around the star. Physical changes in the atmosphere of the star, already underway, will lead to its classification as a so-called Wolf-Rayet star. Wolf-Rayet stars have a high surface temperature, a strong stellar wind and a particular type of peculiar spectrum.

Credit: J. Sandoval (CSC), and NASA

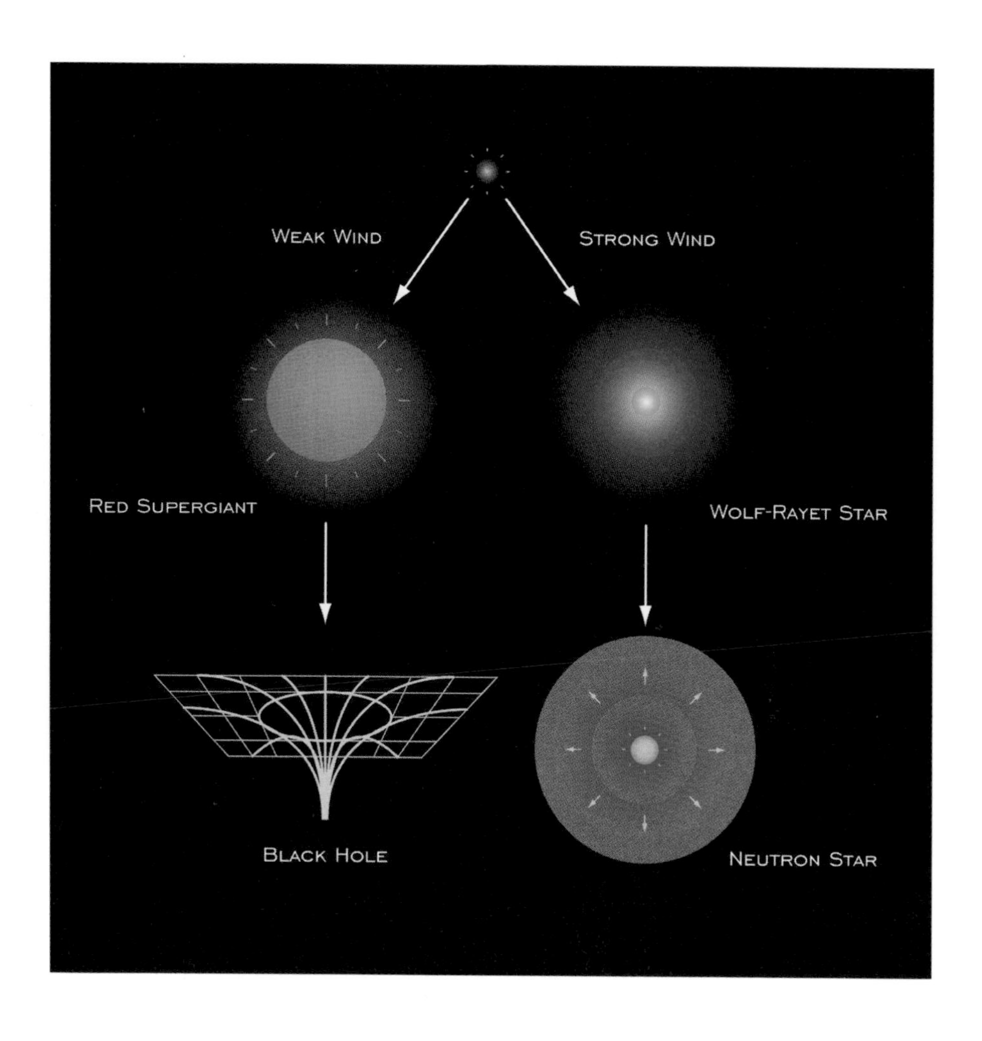
WEAK WIND
STRONG WIND
RED SUPERGIANT
WOLF-RAYET STAR
BLACK HOLE
NEUTRON STAR

Nova Cygni 1992 in May 1993

This is the first of two views of Nova Cygni 1992, an eruption in a binary star system located about 10,000 light years from Earth. The nova was discovered in the constellation Cygnus on February 19, 1992. This image was taken on May 31, 1993. The exploding star has expelled a bubble of gas. It appears brightest around its periphery where we are looking through the greatest thickness of gas. It gives the impression that the formation is like a ring, or hula-hoop, but in reality it is probably shaped more like a squashed ball. A bar of light that appears to run diagonally from upper left to lower right across the shell is unexpected and difficult to explain. This image was taken before the First Servicing Mission in 1994 and it is hard to distinguish fine details close to the bright central star due to the distorting effect of the spherical aberration in the main telescope mirror.

Camera: Faint Object Camera
Credit: F. Paresce (STScI and ESA), NASA, and ESA

Nova Cygni 1992 in January 1994

This image of Nova Cygni 1992 was obtained with the aid of the Faint Object Camera and the corrective optics apparatus (COSTAR) installed on the Hubble Space Telescope during the First Servicing Mission. Compared with the picture taken in May 1993, it is much clearer. It shows that the ring had expanded from a diameter of 118 billion kilometers (74 billion miles) to 154 billion kilometers (96 billion miles) and had become less round. In addition, the mysterious diagonal bar seen in the May 1993 image had vanished.

Camera: Faint Object Camera with COSTAR

Credit: F. Paresce (STScI and ESA), R. Jedrzejewski (STScI), NASA, and ESA

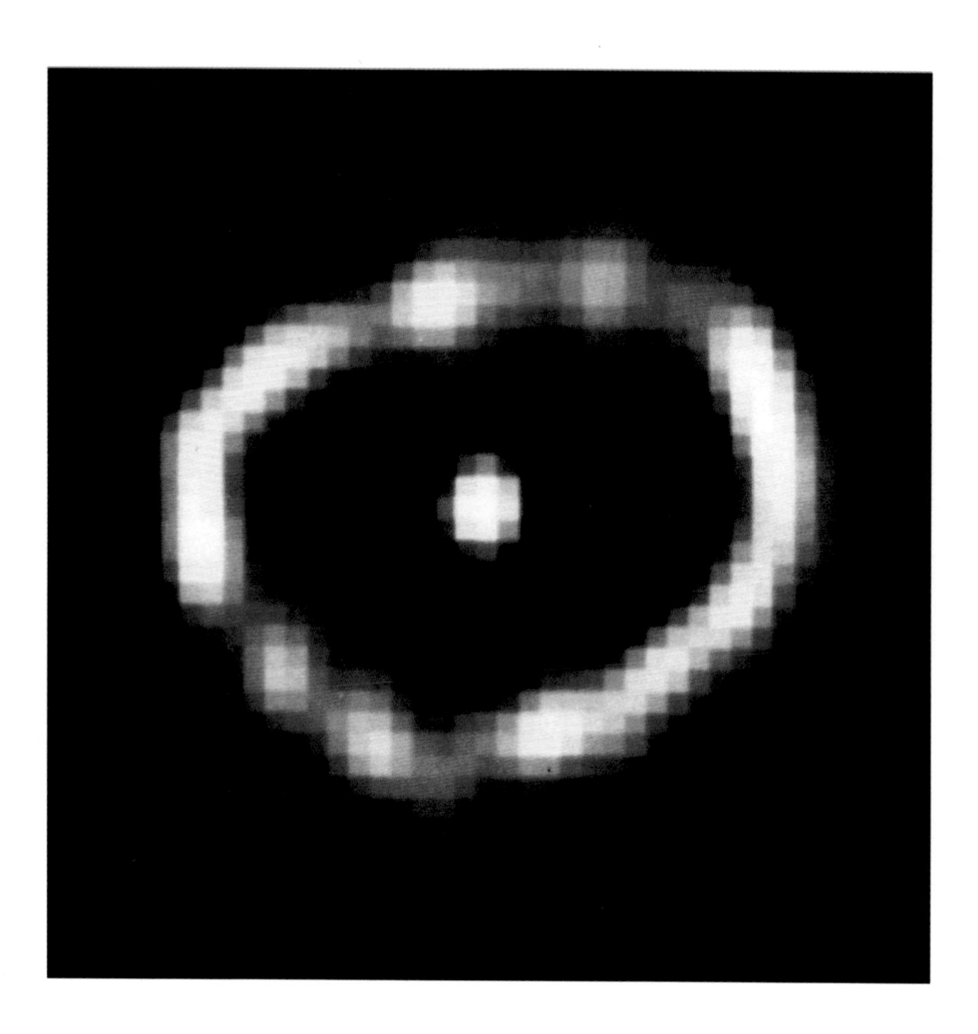

Eta Carinae: Supernova in the Making

Eta Carinae, one of the most massive stars in the Milky Way Galaxy, has been called a 'supernova in the making'. An outburst seen from the southern hemisphere in 1843 made it briefly the second brightest star in the night sky, although it is an estimated 9,000 light years from Earth. The star itself is not seen in this image. It is hidden by an elaborate nebulosity of gas and dust, produced by past eruptions. Yet infrared observations that penetrate the cloak of dust show that it is shining within, and recent observations with the HST have measured the rate at which matter continues to stream from the star.

The larger, red region of nebulosity is probably the most rapidly moving gas that erupted from Eta Carinae in the 1840s. Some of this outlying material is moving at velocities in excess of two million miles per hour. The two pronounced lobes at the center of the picture shine so brightly because they contain huge numbers of microscopic dust particles that reflect or 'scatter' the light from the hidden central star.

Eta Carinae is about 4 million times more luminous than the Sun, and probably more than 100 times as massive. Presumably it will indeed become a supernova within the next few million years.

Camera: WFPC2

Technical Information: Composite made from images taken separately in red, green, and blue light.

Credit: J. Hester (Arizona State University), and NASA

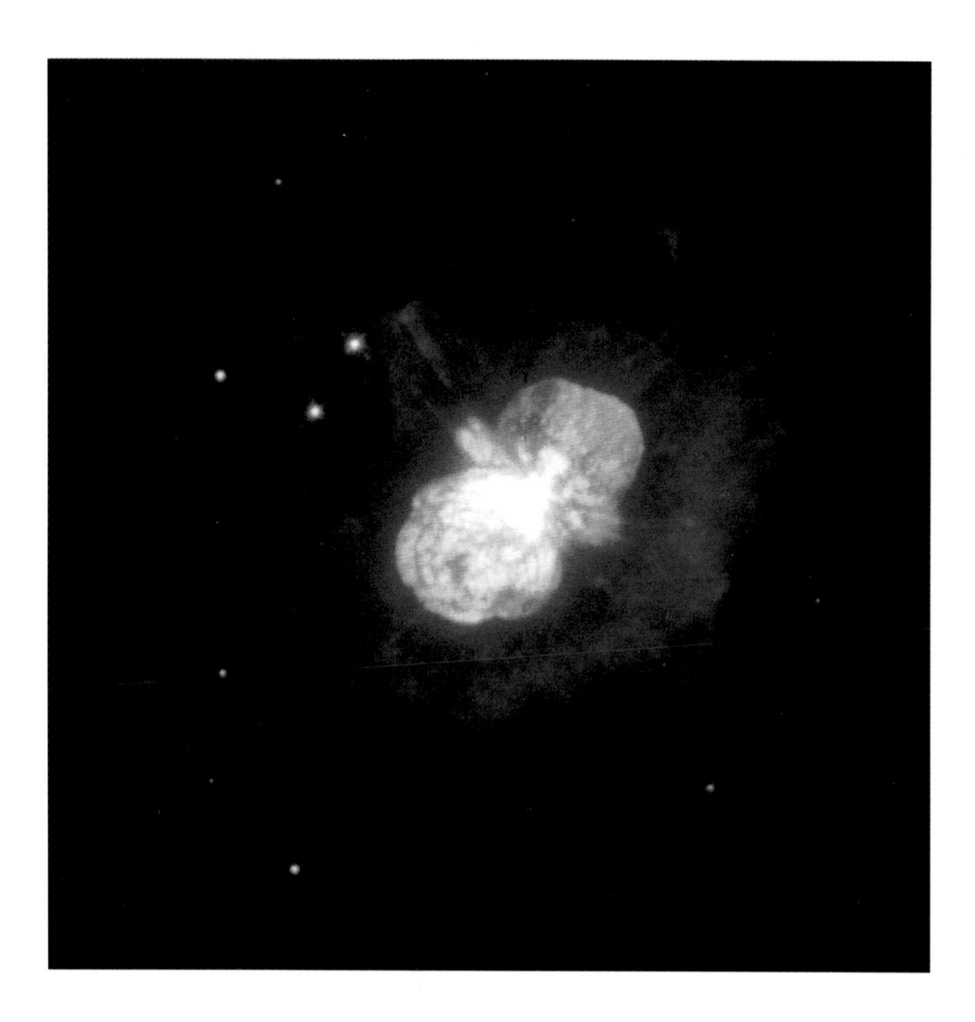

AG Carinae: A Luminous Blue Variable Star

Circumstellar matter surrounding the luminous blue variable star AG Carinae is seen in this HST image. Because the nebulosity is so much fainter than the star, the overexposed central bright image is occulted by the circular black disk. The long vertical white column is an artifact because of the over-exposure in the brightest part of the image. Likewise, the white rayed structure extending from the edge of the black disk is an artifact, not a real stellar feature.

This image is much sharper than previous views through telescopes on the ground. Those had shown the brighter nebulosity extending to the left from AG Carinae as a relatively featureless 'jet'. The 'jet' and other circumstellar nebulosities are here revealed as a complex mix resembling bubbles, filaments, and arches. The nebulosities are gas clouds that contain microscopic dust. They shine because the light from AG Carinae is reflected by the dust particles. The nebulosities are probably shaped by the stellar wind from AG Carinae and swept outwards in its flow.

A luminous blue variable is a supergiant star in an unstable state in which it may change noticeably in brightness and lose much mass. AG Carinae is one of the brightest and most massive stars in the Milky Way, and is located about 21,000 light years from the Sun.

Camera: WFPC2
Credit: A. Nota, M. Clampin, and C. Leitherer (STScI), and NASA

The 'Cat's Eye' Planetary Nebula

NGC 6543, nicknamed the 'Cat's Eye', is one of the most complex planetary nebulae ever seen. It is about 3,000 light years away in the constellation Draco. This HST image reveals concentric shells of gas, high-speed jets and unusual knots of gas. Estimated to be 1,000 years old, the nebula represents a late stage in the evolution of a dying star. The term 'planetary nebula' is a misnomer. Planetary nebulae have nothing to do with planets except that the 18th century observer, Sir William Herschel, thought their appearance in his telescope reminiscent of his views of the planets.

The intricate structures within NGC 6543 are more complicated than normally seen in planetary nebulae and may be explained if there is a double star system at its center. The two stars would be too close to be resolved separately and appear as a single point of light.

A fast stellar wind blowing off one of the central stars created the elongated shell of dense glowing gas. This structure is embedded inside two larger lobes of gas blown off the star earlier. These lobes are pinched by a ring of denser gas, thought to have been ejected in the plane of the orbits the stars follow around each other. The suspected companion might also be responsible for the pair of high-speed jets at right angles to the ring.

Camera: WFPC2

Technical Information: Composite of three images taken in three different colors. Red represents light from hydrogen, blue from neutral oxygen and green from ionized nitrogen.

Credit: J. P. Harrington and K. J. Borkowski (University of Maryland), and NASA

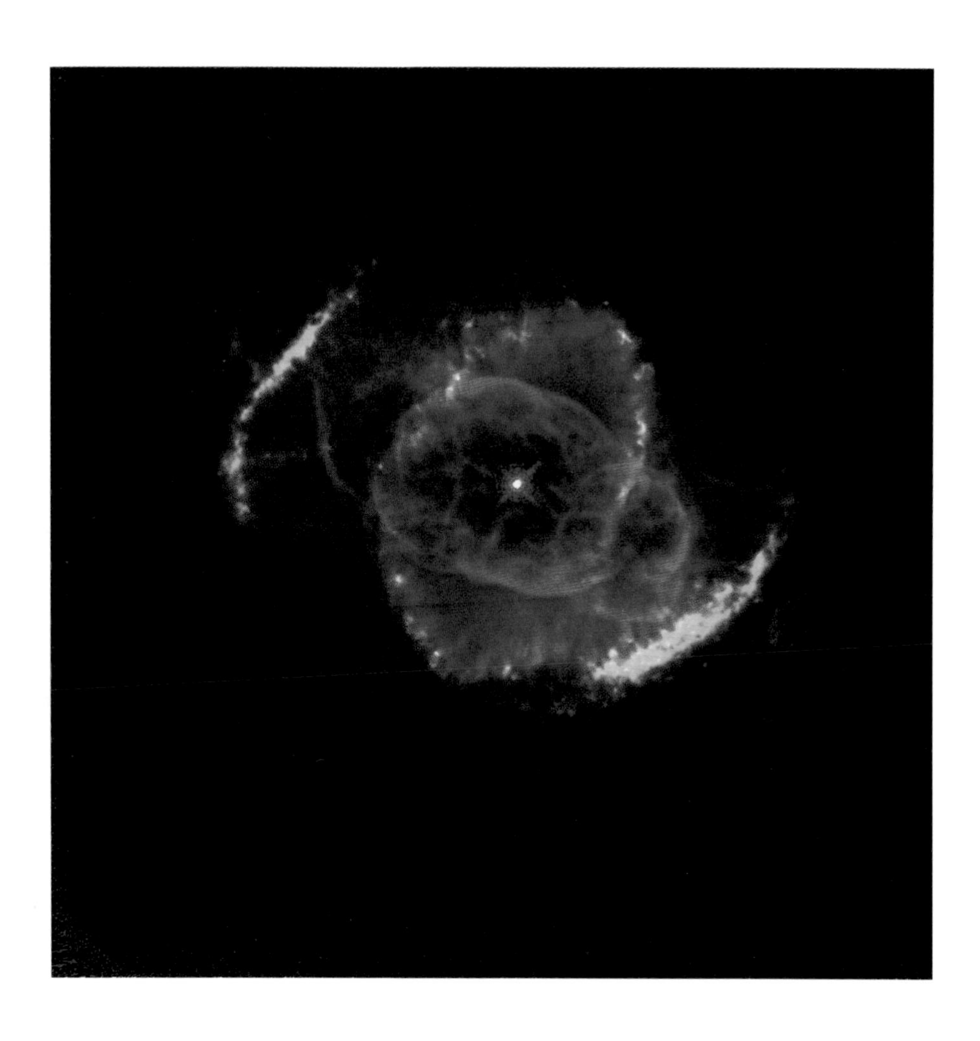

Planetary Nebula N66 in the Large Magellanic Cloud

This planetary nebula, known as N66, was ejected by a luminous red giant star. The star subsequently shrank to create the blue remnant at the center of the nebula. Ultimately it will become a white dwarf. Radiation from this star ionizes the nebula, causing it to glow with both visible and ultraviolet light. The nebula is about 1.9 light years across and individual lobes within it are expanding from the center with speeds of up to one quarter of a million miles per hour (100 kilometers per second).

N66 is located in the Large Magellanic Cloud, a satellite galaxy of the Milky Way, 169,000 light years away.

Camera: Faint Object Camera

Technical Information: Image taken in the light of doubly ionized oxygen at 501 nm and sharpened by computer image reconstruction.

Credit: J. C. Blades (STScI), M. J. Barlow (University College, London), NASA, and ESA

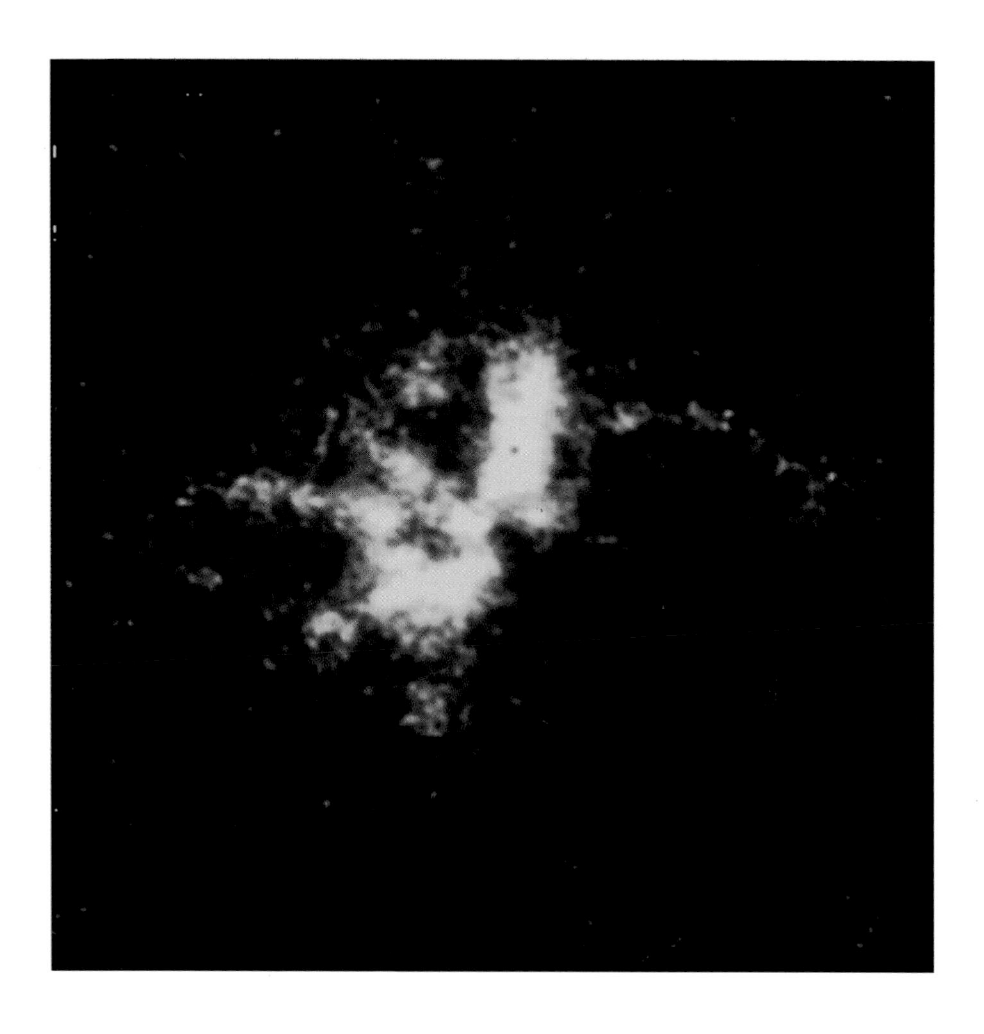

Supernova 1987A in August 1990

Supernova 1987A was the explosive death throes of a blue supergiant star in the Large Magellanic Cloud, a satellite galaxy of the Milky Way. Following the supernova, the star that exploded was identified on earlier telescopic photographs. The supernova was discovered by Ian Shelton at the Las Campanas Observatory, in Chile, on February 24, 1987. Initially visible to the naked eye, it was the brightest supernova since the year 1604.

This false-color Hubble image was taken on August 23, 1990, through a filter that isolates light from doubly ionized oxygen atoms. The debris of the shattered star is the pink blob at the center of the yellow elliptical ring. Comparison of the blob with images of stars taken with the same camera demonstrates that it is really extended, not pointlike, so that the expansion of the stellar debris into surrounding space has been directly imaged. Using HST images of Supernova 1987A and other data, it was possible to measure the distance to the Large Magellanic Cloud more accurately than ever before and it was found to be 169,000 light years away.

Theories abound as to the nature of the elliptical ring, but it is clear that it is too large to represent ejecta from the supernova so soon after the explosion. The ring must have surrounded the blue supergiant before it became a supernova. A flash of ultraviolet light from the exploding star probably excited the gas in the ring and made it glow. Some astrophysicists have concluded that the ring is actually the dense gaseous 'waist' of a thinner, hourglass-shaped formation produced by stellar winds that had blown off the star before it became a supernova.

Camera: Faint Object Camera (1990)
Technical Information: Image taken in the light of doubly ionized oxygen at 501 nm.
Credit: NASA and ESA

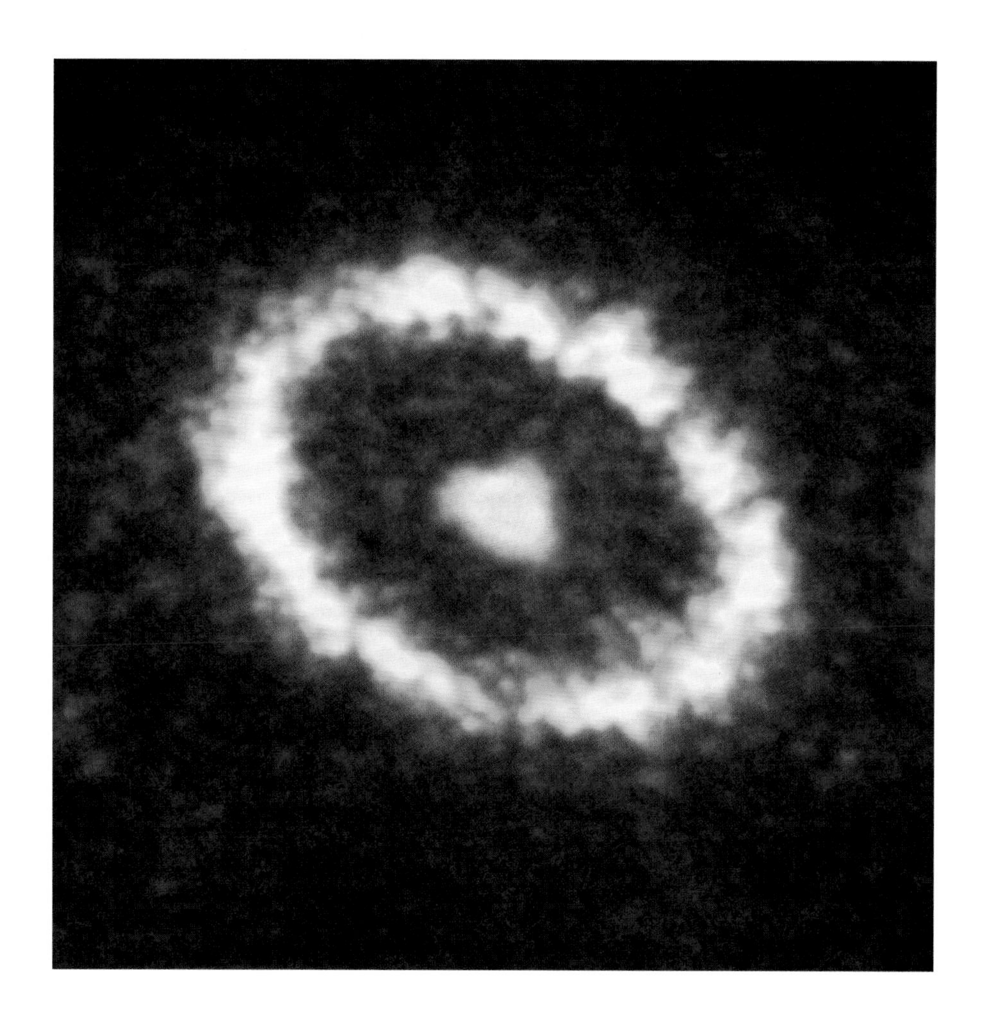

Supernova 1987A in Ultraviolet Light, 1994

This image shows the region around Supernova 1987A as photographed in ultraviolet light on January 8, 1994. Prior to the installation of COSTAR on the HST, the ring around the supernova was not readily discerned in ultraviolet images, if seen at all. But in this image, details of the ring, including its wavy structure, are vividly apparent.

A faint star of about the 20th magnitude seems superimposed on the ring at the lower right. A well-known bright star (which was outside the image area reproduced in the preceding illustration) is partially shown at lower left. The 20th magnitude star has not been studied sufficiently to determine for sure where it is. It may be a rather bright ordinary star located in the Large Magellanic Cloud at approximately the same distance as the supernova. In that case, it appears faint because it is so far away. Alternatively, it may be a small, dim object – a white dwarf star – located much closer to Earth in our own Milky Way Galaxy.

Careful measurements of the debris from Supernova 1987A, which is the bright image centered in the elliptical ring, show that the debris cloud has doubled in size since August 1990. As observed in January 1994, the cloud image had a diameter of one-quarter of a light year. Thinner, faster-moving gas on the outskirts of the debris cloud may actually have reached the ring by now. The main debris should strike the ring at some time around the year 2000, give or take a few years. When that happens, a strong burst of x-rays is likely to occur, and astronomers will use the HST to try to determine the effect of the collision on the ring.

Camera: Faint Object Camera with COSTAR

Credit: P. Jakobsen (ESA/ESTEC), F. D. Macchetto (STScI and ESA), R. Jedrzejewski (STScI), N Panagia (STScI and ESA), NASA, and ESA

Supernova 1987A's Outer Rings

Subsequent to the discovery of the bright elliptical ring around Supernova 1987A, which is shown in the preceding two pictures, astronomers found two much larger and dimmer nebulous structures located to either side of the supernova remnant. Their nature was in doubt until this remarkable image was obtained with the HST in February 1994. The image, taken in the red light of the hydrogen atom, shows that the two dimmer structures are also rings. Furthermore, it seems that the two outer rings are hoop-shaped but are so narrow that even the HST cannot determine how wide any given part of the hoop is.

The nature of the two outer rings near Supernova 1987A is the subject of much conjecture. Are they real nebulosities or are they emission patterns, 'painted' on a dark canvas by beams of energetic particles or radiation from a strategically located object? Or yet something else? It is fair to say that they have not been explained to the general satisfaction of astronomers. This is not very surprising, as no similar phenomenon has ever been seen in space.

Camera: WFPC2

Credit: C. Burrows (STScI and ESA), and NASA

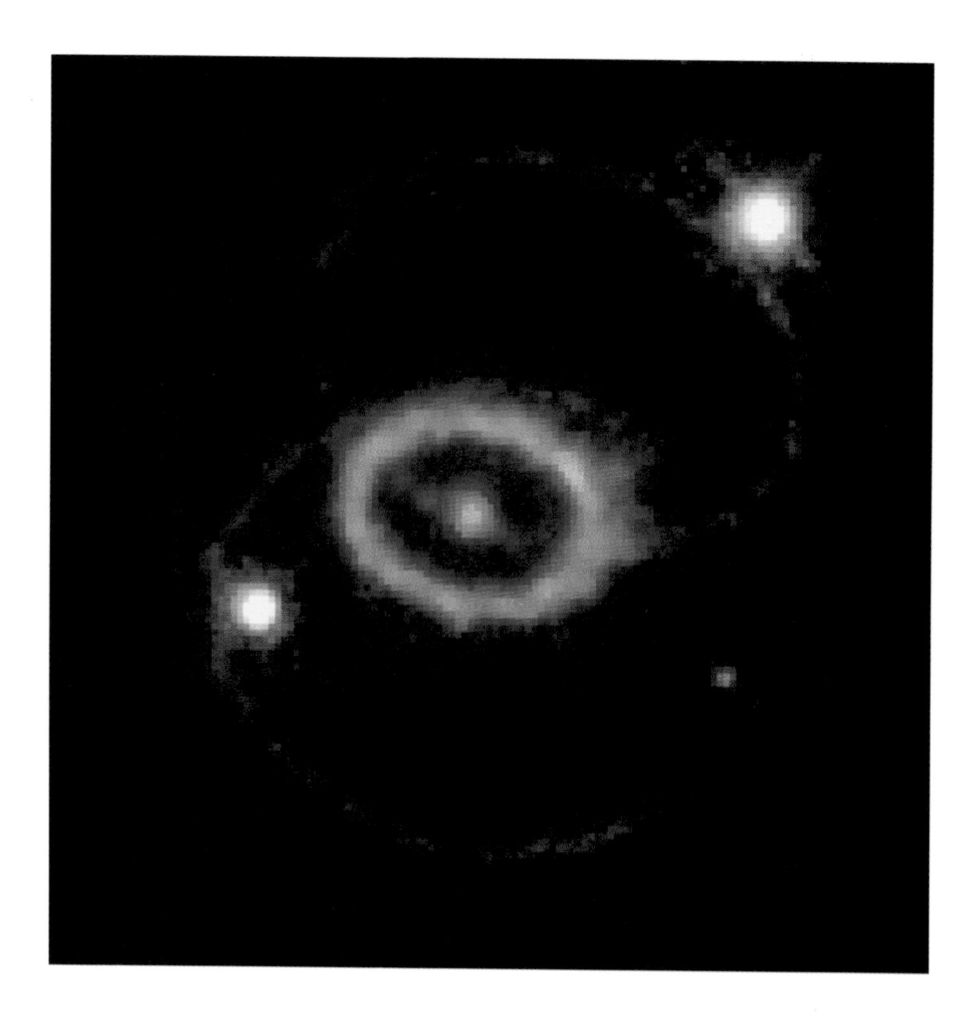

Filaments in the Crab Nebula

The Crab Nebula was produced in a supernova explosion that was witnessed in Asia and perhaps also in North America and elsewhere, on or about July 4, 1054 AD. It is located about 6,000 light years from Earth. The filaments in the nebula are made of material from the outer regions of the exploded star. A thinner and more diffuse gas cloud (not seen in these images) consists of electrons and positrons moving outwards at nearly the speed of light. These are accelerated from an ongoing source, identified as the pulsar at the heart of the Crab Nebula. The pulsar is a very compact and rapidly spinning neutron star, the imploded remnant of the core of massive star that blew up as the supernova.

This image is a false-color composite in which the colors indicate regions of different temperatures, with the coolest gas shown in red, warmer gas in blue, and the hottest gas in green.

Camera: WFPC2

Technical Information: Composite of three images taken in the light of cool oxygen (red), sulfur (blue), and hot oxygen (green).

Credit: J. Hester and P. Scowen (Arizona State University), and NASA

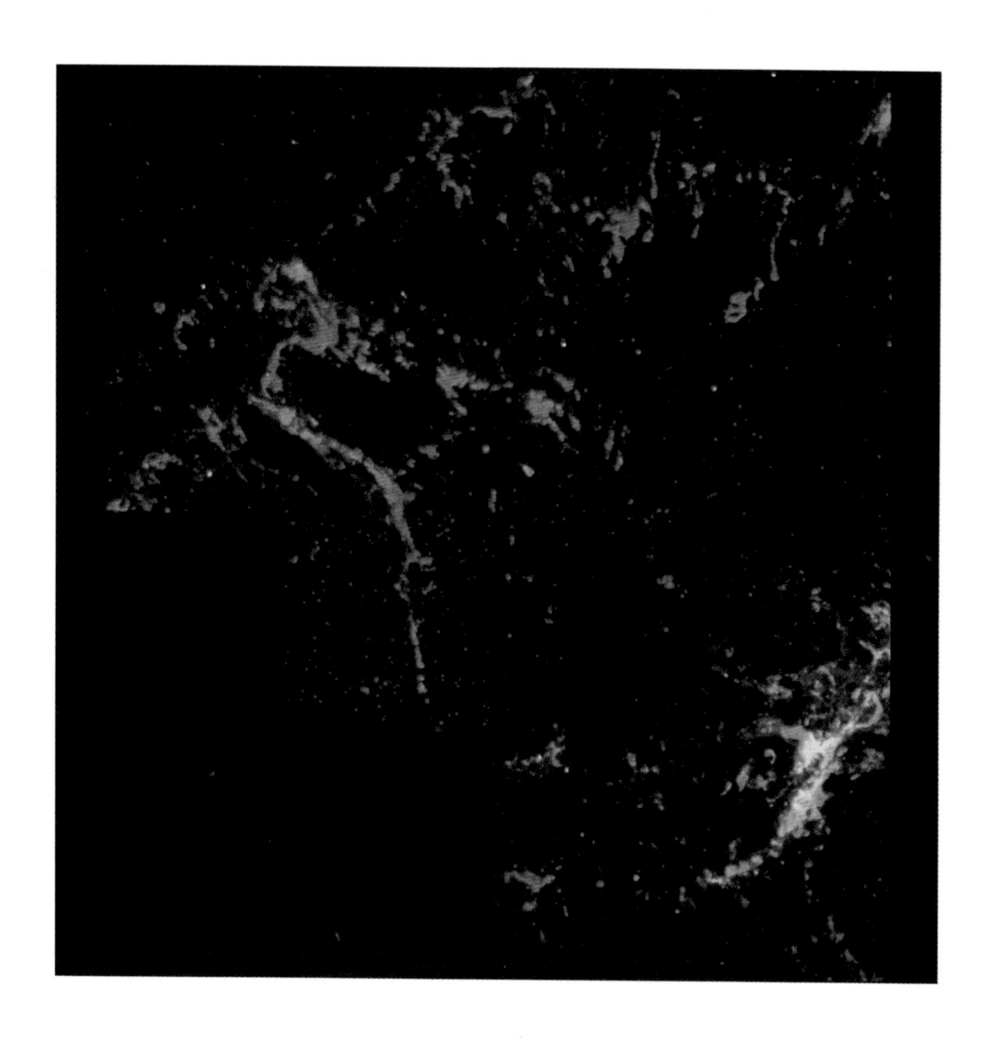

Details of Filaments in the Crab Nebula

This small region of the Crab Nebula is an enlargement of part of the preceding picture. Due to the filters employed, the image shows the light from the gaseous filaments of the nebula – shreds of the shattered supernova star – and does not show the thinner and hotter part of the nebula that consists of electrons and protons streaming outward at high velocity from the pulsar at the heart of the Crab Nebula. However, the picture suggests the likely nature of the process that gives many of the filaments their shape. Seen at the high resolution of the WFPC2, each small filament has a head consisting of a dense clump of gas, rather like the head of a comet (although a comet also has a small solid nucleus within its head and these filaments do not). Stretching back from the head, the filament may be like a streamer of gas blown outward from the head by the thin hot gas, which is expanding outward from the pulsar.

Camera: WFPC2

Technical information: Composite of three images taken in the light of cool oxygen (red), sulfur (blue), and hot oxygen (green).

Credit: J. Hester and P. Scowen (Arizona State University), and NASA

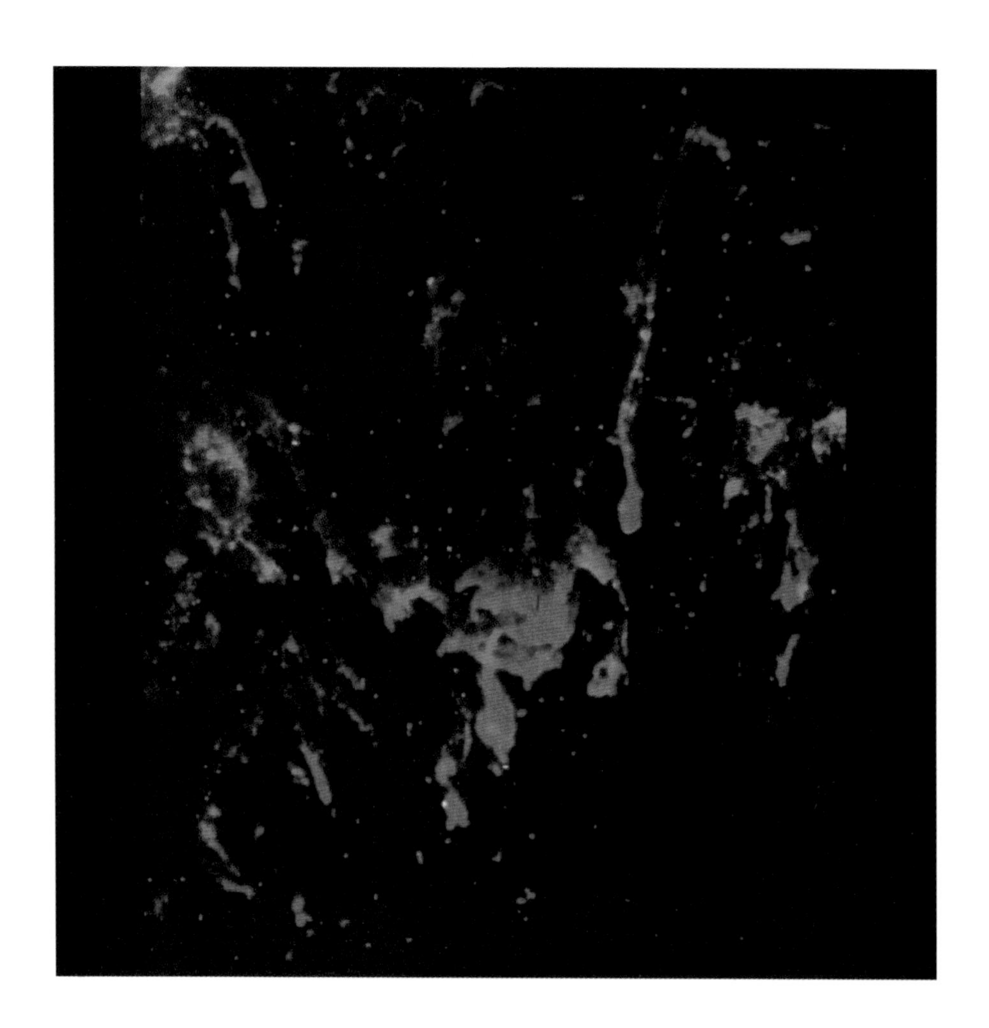

The Pulsar and the Wisps in the Crab Nebula

This picture shows a nebulous 'knot' next to the pulsar in the Crab Nebula. The pulsar is the brighter object of the close pair at the center of this image. The dimmer object next to the pulsar is the 'knot' of nebulosity. It is actually about 225 billion kilometers (140 billion miles, or 1,500 times the distance from the Earth to the Sun) from the pulsar. The two objects appear to be close to one another because they are being viewed at the distance of the Crab Nebula, about 6,000 light years from the Earth.

The knot may have been present for a long time, or it may be a new phenomenon. Previous telescopic observations were incapable of distinguishing it. According to one theory, the knot may be the location where a high-speed wind of electrons and positrons streaming outward from the pulsar undergoes a shock wave, resulting in material piling up like snow on the blade of a plow. The line from the pulsar to the knot lies along the direction of a jet structure of uncertain nature that can be seen in x-ray images of this region. This jet may extend from the polar cap of the pulsar.

The wispy structures seen in this image are not shreds of the exploded supernova star, but a less well-understood phenomenon that appears to be directly associated with the pulsar. Time sequences of photographs made in past years with telescopes on the ground, while much less sharp than this HST image, indicate that these pulsar wisps move, or that they appear and disappear, or both.

Camera: WFPC2
Credit: J. Hester and P. Scowen (Arizona State University), and NASA

Understanding Features near the Crab Nebula Pulsar

This diagram gives an explanation of the features seen in the previous image. The distance scale is shown by the bar at the lower left, which represents 10,000 astronomical units. One astronomical unit, approximately the distance between Earth and the Sun, is equivalent to 150 million kilometers or 93 million miles.

At the center of the image is the pulsar itself and the newly discovered knot of emission. The two largest arrows indicate the polar axis of the pulsar. A jet of x-ray emission has been detected (in other observations) along this direction. The wisps seen above and to the right of the pulsar may form a ring-like 'halo' (the dashed oval shape), which is foreshortened because it is tipped with respect to our line of sight. It is thought that a polar jet of pulsar wind streams through the center of this halo, and that the ring may represent the interface or boundary between this stream and another wind that blows outwards from the equatorial region of the pulsar.

Credit: J. Hester and P. Scowen (Arizona State University), and NASA

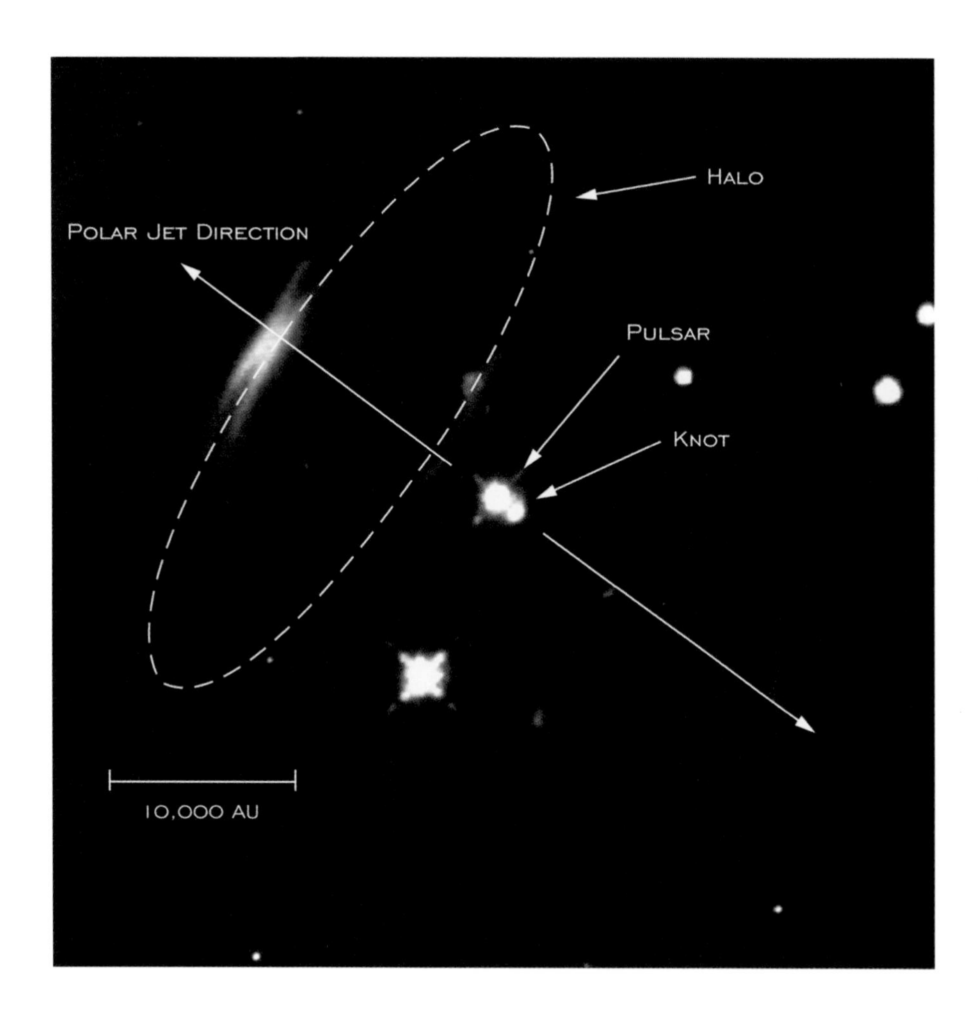
HALO
POLAR JET DIRECTION
PULSAR
KNOT
10,000 AU

The Cygnus Loop Supernova Remnant

The Cygnus Loop (also called the Veil Nebula), is a dim ring of glowing gas invisible to the unaided eye, which extends over a region six times the apparent diameter of the Full Moon in the northern constellation Cygnus. It lies about 2,500 light years away. Although listed as a supernova remnant, the remains of an exploded star, the Cygnus Loop actually consists overwhelmingly of interstellar clouds that have been swept up by the blast wave from the supernova explosion that occurred an estimated 15,000 years ago. The blast wave has heated the interstellar clouds to temperatures of tens of thousands of degrees. In this view, the blast wave is moving from left to right and has recently hit a cloud of interstellar gas that is denser than average, causing it to glow.

The image shown here is a false color composite of three images, taken through filters that isolate the light from three kinds of atoms, each representing a different temperature range in the Cygnus Loop. The light from oxygen ions (with temperatures of 30,000 to 60,000 kelvins, or about 50,000 to 100,000 degrees Fahrenheit) is shown in blue. The light from hydrogen atoms that occur throughout the Loop is represented in green. Much of the hydrogen emission comes from an extremely thin zone immediately behind the shock front itself. These thin regions appear as sharp green filaments in the image. The light from sulfur atoms, shown in red, is produced in gas that has cooled to about 10,000 kelvins (18,000 degrees Fahrenheit).

Camera: WFPC2

Credit: J. Hester (Arizona State University), and NASA

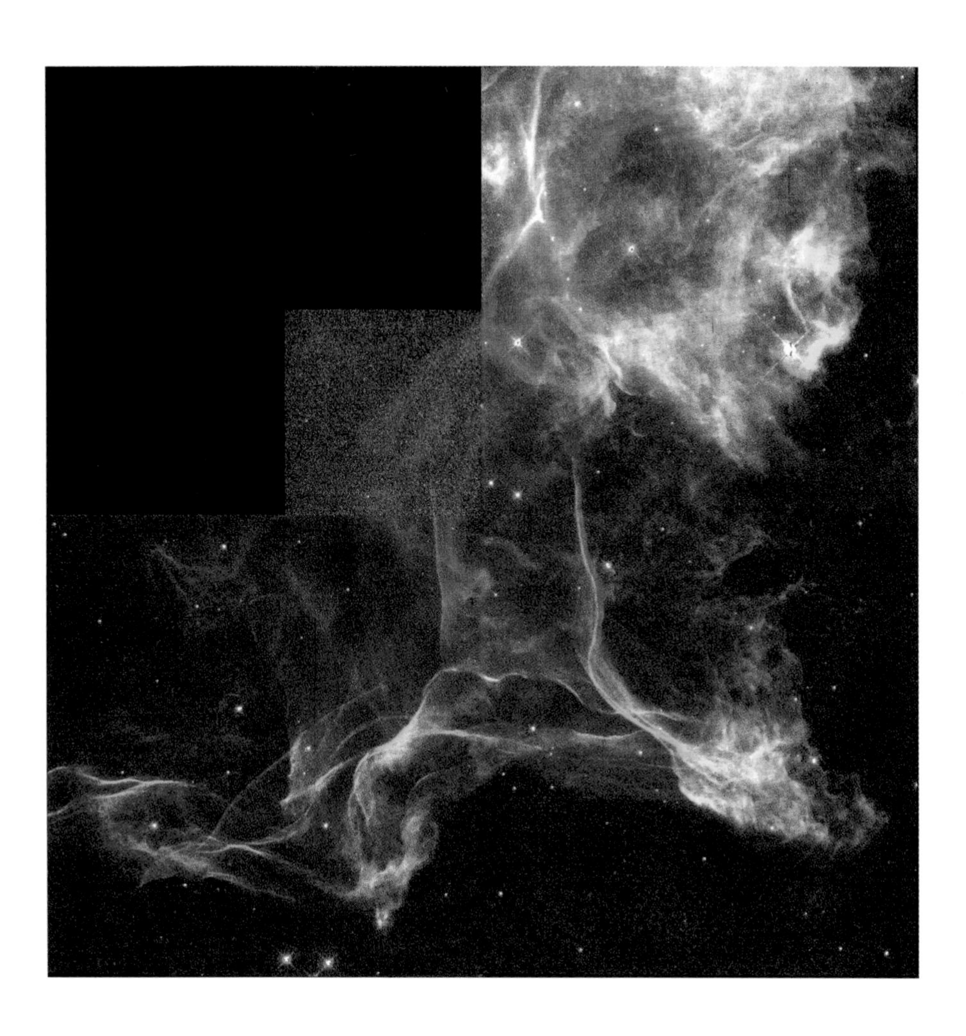

The Center of a Giant Elliptical Galaxy, NGC 1275

NGC 1275, one of the most luminous objects in the local universe, is the central galaxy of the Perseus Cluster of galaxies, which is located about 250 million light years from Earth. Only a small part of the galaxy is covered by this image. The bright white spot is the nucleus or center of NGC 1275. The blue dots are bright, dense concentrations of stars, possibly young globular star clusters. If so, each of these little dots is actually a densely packed collection of hundreds of thousands or more stars, like 47 Tucanae and M15 in our own Galaxy. However, the stars in them would be much younger than those in globular clusters of the Milky Way. Globular star clusters are among the oldest objects in the Milky Way, and are composed exclusively of very old stars.

NGC 1275, the central and largest galaxy in the Perseus Cluster, has probably merged with several smaller galaxies. When galaxies merge or make close passes, few if any of their billions of stars actually smash into each other because the spaces between the stars are just too wide. But gravitational forces can wreak havoc on the shape of galaxies, and gas clouds, which are much larger than stars, do collide. Theorists suggest that in the case of NGC 1275, the merger process may have triggered the collapse of massive interstellar gas clouds to form the objects identified as young globular clusters.

Camera: WFPC2
Credit: J. Holtzman (UCSC), and NASA

The Cartwheel Galaxy

The Cartwheel Galaxy, 500 million light years away in the constellation Sculptor, is the spectacular result of a head-on collision between two galaxies. It was presumably a normal spiral galaxy, similar to the Milky Way, before a small intruder careered through its very center. One of the two small galaxies to the right of the Cartwheel itself may be the culprit, but it is not clear which. The small blue galaxy is disrupted and has new star formation, which strongly suggests it is the interloper. However, the smoother looking yellowish galaxy has no gas, which is consistent with the idea that gas was stripped out of it during passage through the Cartwheel Galaxy.

The collision triggered a ripple of energy through the Cartwheel that is plowing gas and dust in front of it. Expanding at 200,000 miles per hour, this 'cosmic tsunami' leaves bursts of new star formation in its wake. The bright blue knots in the rim of the 'wheel' are gigantic clusters of newborn stars. The ring is so large – 150,000 light years across – that the whole of the Milky Way would fit inside it. It contains at least several billion new stars that would not normally have been created in such a short time span. This HST image also resolves immense loops and bubbles blown off by exploding supernovae. The faint 'spokes' between the 'rim' and the 'hub' suggest that spiral structure is beginning to re-emerge.

Camera: WFPC2
Technical Information: Composite of two images taken in blue and near-infrared light.
Credit: K. Borne (STScI), and NASA

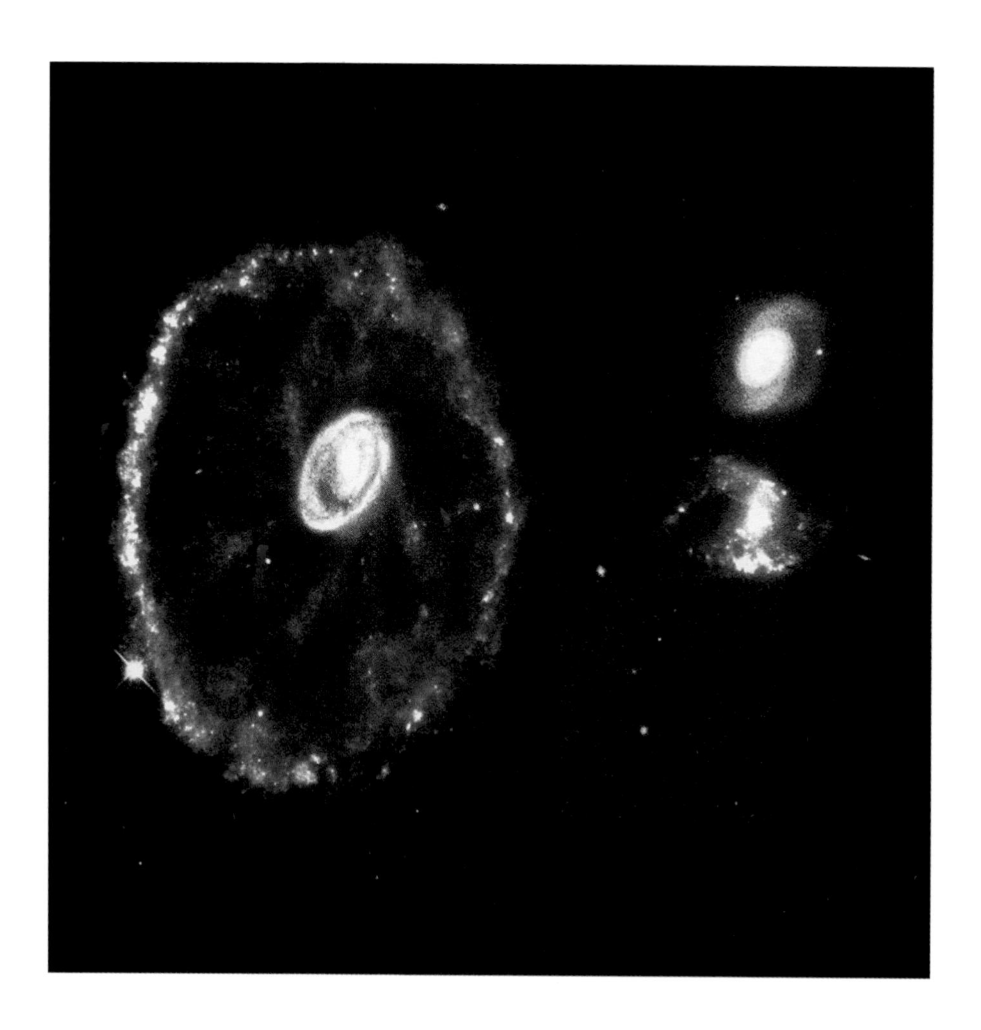

Colliding Galaxies, NGC 7252

This image shows the center of the colliding galaxies system NGC 7252. Located about 300 million light years from Earth in the direction of the constellation Aquarius, it appears to be a clear example of two disk galaxies in the process of merging after a collision. The orderly spiral pattern is in peaceful contrast to the chaotic outer regions of NGC 7252, as shown in the inset picture at the upper left, a ground-based image from the 4-meter telescope at Cerro Tololo Inter-American Observatory, in Chile. The two long plumes, consisting of stars and gas ripped out of the merging galaxies by gravitational force, are called 'tidal tails.'

The bright dots in this HST image of NGC 7252 are resolved by the telescope. Each appears about 0.04 arc seconds across, the apparent size of a small coin seen from a distance of 100 kilometers (60 miles). Their real diameters are about 60 light years, similar to the sizes of globular clusters in the Milky Way. Researchers have concluded that they are young globular star clusters, formed in the collision of the two disk galaxies that began about one billion years ago. This conclusion is supported by spectrograms obtained with the 5-meter (200-inch) Hale Telescope at Palomar Observatory.

A long-standing theory proposes that elliptical galaxies are formed from collisions between disk galaxies. An objection to the theory is that elliptical galaxies of a given mass tend to contain many more globular clusters than comparable disk galaxies. However, the HST findings suggest that the galaxy collisions themselves may produce new globular clusters, perhaps answering this objection. It appears that the small spiral pattern at the center of NGC 7252 is a relatively short-lived phenomenon and that NGC 7252 is well on the way to becoming an elliptical galaxy.

Camera: WF/PC-1 (1992)
Technical Information: False color image with computer image reconstruction.
Credit (HST image): B. Whitmore (STScI), and NASA
Credit (inset): F. Schweizer (Carnegie Institution of Washington)

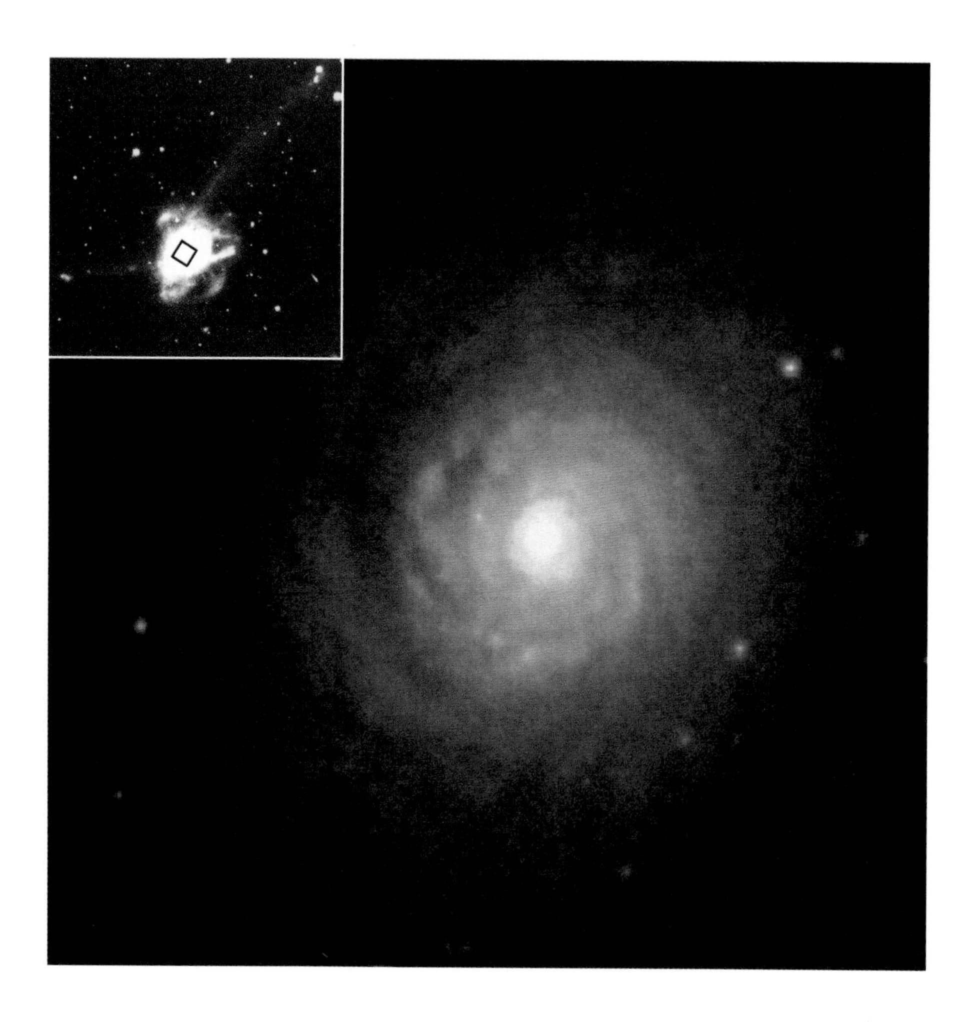

Markarian 315, a Seyfert Galaxy

Seyfert galaxies are spiral galaxies with very bright nuclei and spectra that are often reminiscent of quasars, although the nuclei of Seyferts are not as bright as quasars. They were first described by Carl Seyfert in 1943. As in the case of quasars, astronomers believe there may be a supermassive black hole at the center of a Seyfert galaxy nucleus.

In this image of the central region of the Seyfert galaxy Markarian 315, the brighter of the two bright spots is the nucleus of the galaxy. The smaller spot was discovered in the HST observations. It is separated from the nucleus by about two arc seconds on the sky, equivalent to a real distance of about 6,000 light years. The smaller spot may represent the central core of a now-vanished galaxy that collided with Markarian 315 and was swallowed up by it.

Wider-angle observations of Markarian 315 obtained with telescopes on the ground show a plume of gas about 240,000 light years long. The plume may be a 'tidal tail', ripped out of the colliding galaxies as they merged. Other gas disturbed by the collision may be falling toward the suspected black hole in the nucleus of Markarian 315, being heated by friction as it does so and thereby making the nuclear region shine brightly.

Camera: WF/PC-1
Technical Information: Image shown in false color.
Credit: J. MacKenty (STScI), and NASA

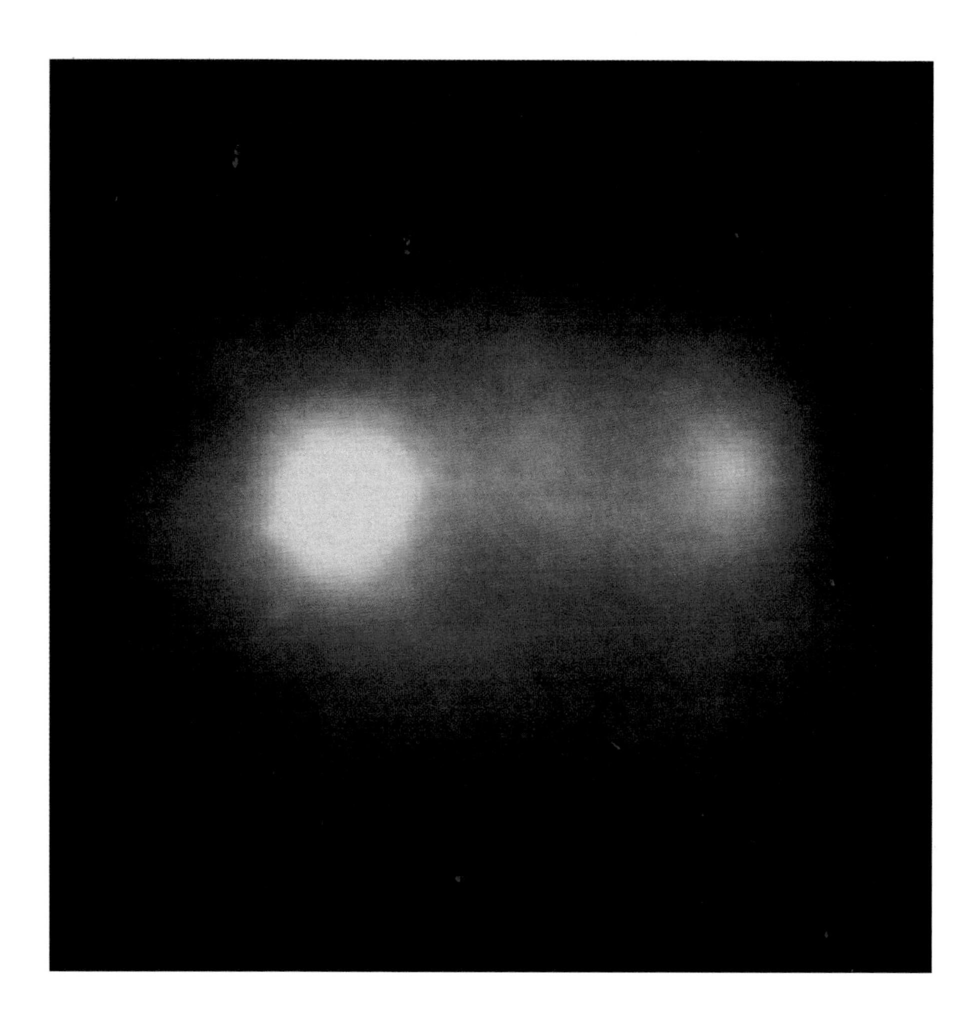

The Quasar PKS 2349

Quasars have remained enigmatic ever since their discovery in 1963 because of the way they emit prodigious amounts of energy yet are very compact. Their star-like appearance gave birth to the name quasi-stellar object, later shortened to 'quasar'. The most widely accepted theory to explain quasars is that they are powered by supermassive black holes in the cores of galaxies.

This HST image of the quasar PKS 2349 shows evidence that it is merging with a companion galaxy. The bright central object is the quasar itself, which is several billion light years away. The wisps next to it are remnants of a bright galaxy that has been disrupted by the gravitational pull between it and the quasar.

Camera: WFPC2

Credit: J. Bahcall (Institute for Advanced Study, Princeton), and NASA

The Center of the Giant Elliptical Galaxy M87

Messier 87 (M87) is a giant elliptical galaxy in the Virgo Cluster of galaxies, roughly 50 million light years from Earth. The HST has peered into its center to seek evidence for the presence of a supermassive black hole.

This near-infrared image of the center of M87 shows a jet of ionized gas or plasma extending toward the right from an extremely bright spot. The bright spot is unresolved, meaning that it must be less than about 6.5 light years in diameter. It does not appear to be starlight, and may be produced by radiation from high-speed electrons moving through a magnetic field. Such electrons also produce powerful radio emission. M87 is the brightest radio source in the constellation Virgo, known to radio astronomers as Virgo A.

The jet, extending more than 5,000 light years from the center of M87 in the HST images, is believed to consist of material expelled from the vicinity of a supermassive black hole in the center of the galaxy. The general background of starlight in M87 increases systematically toward the center in a way that is consistent with mathematical predictions, on the assumption that the concentration of stars is under the gravitational influence of a central black hole with 3.6 billion times the mass of the Sun. The density of stars near the center appears to be over 1,000 times greater than in the vicinity of the Earth, and could be much larger.

Also visible in the image are a number of star-like points of lights scattered about the center. These are globular star clusters within M87, each composed of 100 thousand to a million stars.

Camera: WF/PC-1
Technical Information: Near infrared image at wavelength of 890 nm.
Credit: T. Lauer (NOAO), S. Faber (UCSC), and NASA

The Jet of M87 in Ultraviolet Light

This ultraviolet image of the jet at the center of galaxy M87 shows it consists of a string of knots along a narrow cone that opens out in the direction away from the center of M87. Features as small as ten light years in diameter can be seen.

The bright knots in the jet are possibly regions where the high-speed electrons are subject to additional acceleration.

In this view, the background light of billions of stars in M87 is naturally suppressed relative to the jet. This occurs because the jet shines brightly in the ultraviolet, while the stars, which are old objects and mostly yellow, orange, or red, produce little ultraviolet light.

Camera: Faint Object Camera
Credit: NASA and ESA

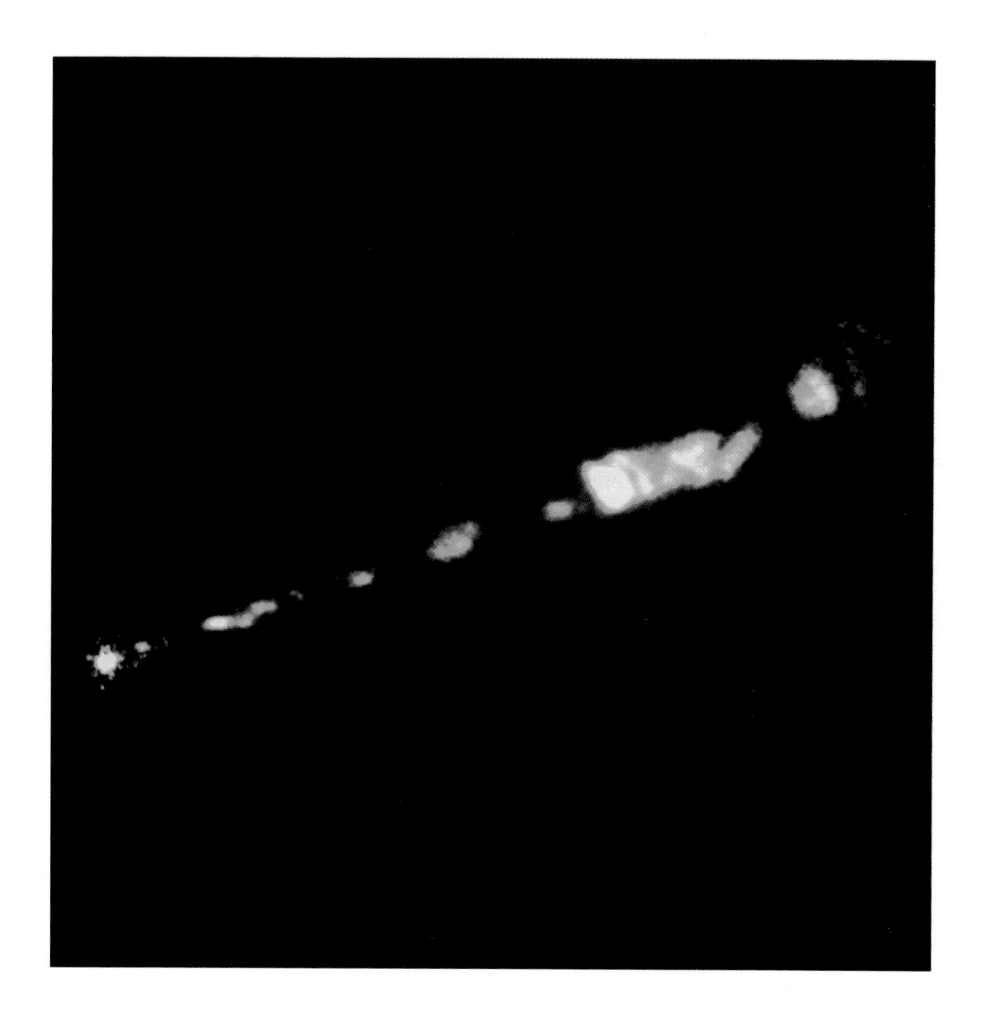

The Gas Disk at the Heart of M87

This visible-light picture of the central region of galaxy M87 is a specially prepared composite. One image was taken in red light, filtered to emphasize radiation from interstellar gas in the form of hydrogen and singly ionized nitrogen (nitrogen atoms that have each lost one electron). A second image was taken at a wavelength where the radiation from M87 is dominated by the glow from the millions of stars clustered toward the center of the galaxy. Then, the second image, adjusted by an appropriate weighting factor, was subtracted from the first one. The result is a composite picture in which the light from interstellar gas shines brightly and the light from stars is de-emphasized.

The result reveals a previously unknown disk of glowing gas surrounding the very center of M87, which is at the base of the jet that extends toward the right corner of the image. A spiral structure is apparent in the gas disk. The gas has a temperature of roughly 10,000 kelvins. It is extremely unusual to observe spiral structure at the center of an elliptical galaxy.

Camera: WFPC2
Credit: H. Ford (STScI and Johns Hopkins University), and NASA

Evidence for a Black Hole in M87

This diagram presents the strong evidence for the existence of a supermassive black hole at the center of galaxy M87. The blue and red tracings represent the spectra of the light from the two locations marked by the blue and red circles on the inset image of the gaseous disk in M87.

The vertical axis of the graph is the intensity of light and the horizontal axis is wavelength. The horizontal displacement on this diagram between the blue and red tracings is due to the Doppler effect. The wavelength of light coming from a source that is approaching the observer is systematically blue-shifted to shorter wavelengths (to the left in this diagram), while light coming from a receding source is red-shifted to longer wavelengths (to the right).

The displacement between the two spectra indicates that, at 60 light years from the center of the galaxy where the observations were made, the gas disk in M87 is turning at the enormous rate of 1.2 million miles per hour. Without the presence of an immense inward gravitational attraction, a disk of gas rotating this rapidly would fly apart. Additional HST observations confirmed that the disk is rotating this rapidly, and that the speed of rotation increases toward the center. The disk is tilted with respect to the line of sight from Earth. As seen here, the disk is turning in the clockwise direction.

The gas disk's huge rotation speed is consistent with the presence of a central black hole of close to 3 billion times the mass of the Sun. Radio astronomy observations have shown previously that the size of the central energy source in M87, at the base of the jet, is a tiny fraction of this 60 light year radius. The combination of enormous mass and very small size implicates a supermassive black hole as the central object of M87. There may be alternate explanations, but they are even more extraordinary and unlikely.

Instrument: Faint Object Spectrograph

Credit: H. Ford (STScI and Johns Hopkins University), and NASA

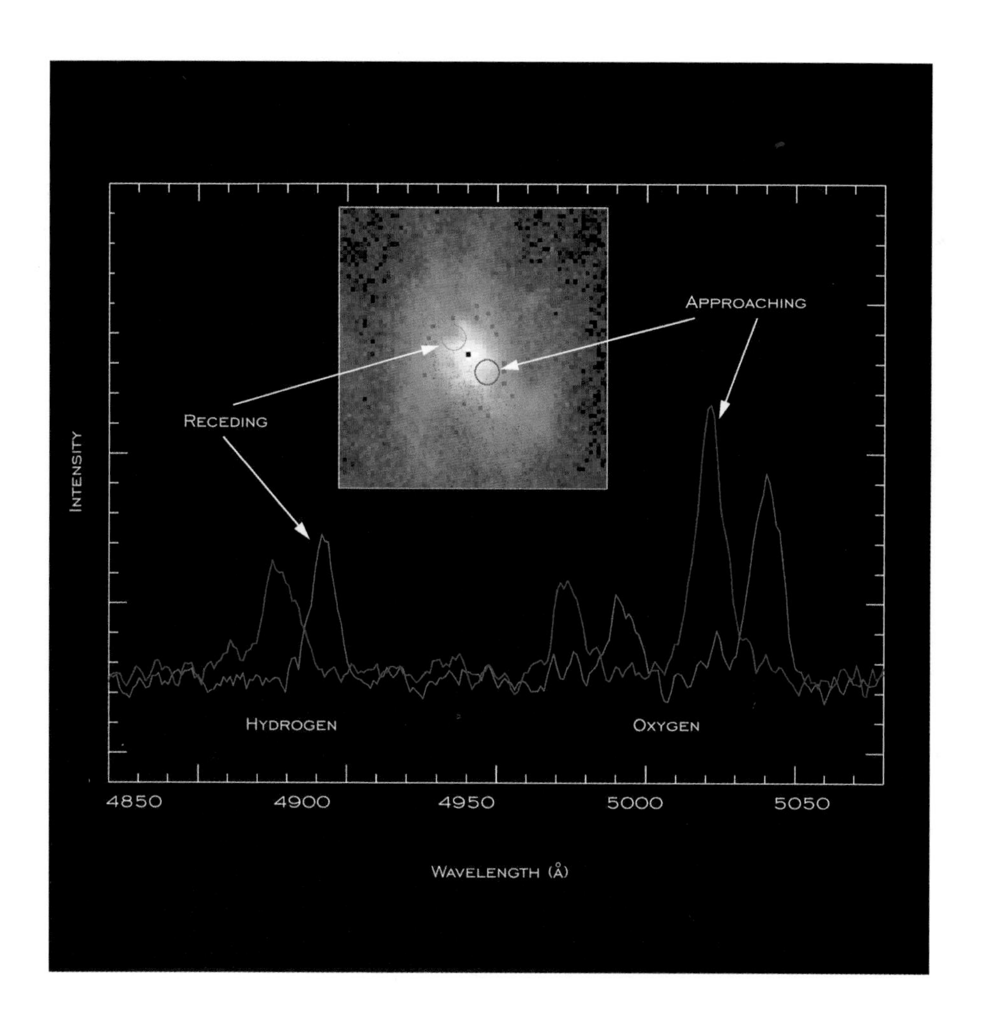

Approaching
Receding
Intensity
Hydrogen
Oxygen
4850
4900
4950
5000
5050
Wavelength (Å)

A Black Hole in the Giant Elliptical Galaxy NGC 4261

NGC 4261 is one of the dozen brightest galaxies in the Virgo Cluster of galaxies and lies about 100 million light years away. This HST image of the heart of the galaxy reveals a remarkable spiral-shaped disk of dust, about 800 light years across, which is 'feeding' a massive black hole at its center. From the speed of the gas swirling around it, astronomers calculate that the black hole's mass is 1.2 billion times that of the Sun, yet all that material is concentrated in a region of space not much larger than our solar system.

Before HST observations, astronomers did not think dust was common in elliptical galaxies such as NGC 4261. Elliptical galaxies were thought to have stopped making stars long ago because there was no gas and dust left in them. However, it now seems that dust and disks are after all present in the centers of these galaxies. The disk in NGC 4261 could be explained as the remnant of a small galaxy that fell into the larger one. Collisions between galaxies were probably more common in the past when the universe was smaller and galaxies closer together. Researchers predict that the black hole in NGC 4261 will swallow the remains of the intruder over the next 100 million years.

A puzzling feature of NGC 4261 is the fact that the black hole is offset by about 20 light years from the precise center of the galaxy and the disk. One exotic explanation suggests that the black hole is self-propelled. Jets of hot gas blasted out from the vicinity of the black hole are observed by radio telescopes as twin lobes of radio emission extending far beyond the visible galaxy. These jets may be pushing the black hole along, rather like a rocket engine.

Camera: WFPC2

Credit: H. Ford (ST ScI and Johns Hopkins University), L. Ferrarese (ST ScI), and NASA

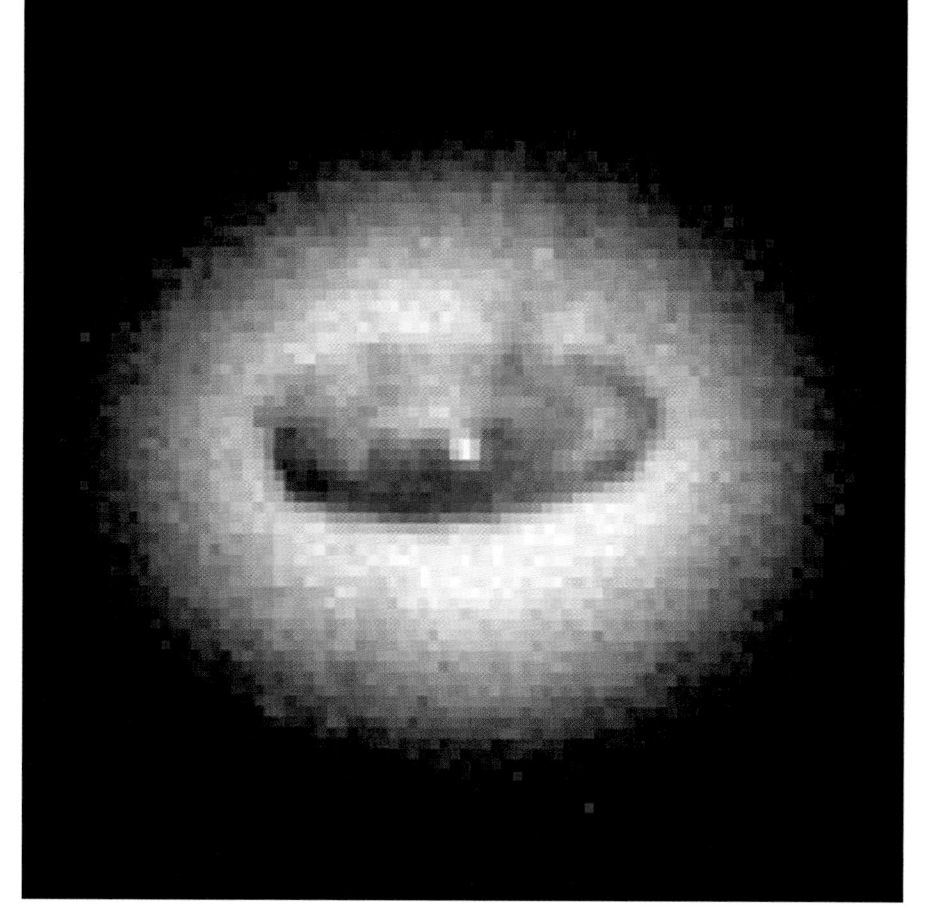

The Einstein Cross

As predicted by Einstein's General Theory of Relativity, the path light takes is deflected measurably from a straight line when it passes close to a massive object. As a result, a galaxy can act in much the same way as an irregular glass lens, to produce a distorted or multiple image of anything that happens to lie behind it along the same line of sight. A single quasar may appear as two or more quasars. And a distant galaxy may be distorted almost beyond recognition. The phenomenon is called 'gravitational lensing'.

The four outer bright spots in this gravitationally lensed system, known as the Einstein Cross (or by its catalog number G2237+0305), are all images of a single distant quasar. The bright spot in the center is a direct image of the central bulge of a spiral galaxy located roughly 400 million light years from Earth. The spiral galaxy is acting as the gravitational lens. Its gravity bends the light from the single, more distant quasar into the pattern of four that we see from Earth. The quasar is about 20 times farther from Earth than the lensing galaxy.

The Einstein Cross was discovered by John Huchra of the Harvard-Smithsonian Center for Astrophysics at the Whipple Observatory near Amado, Arizona, USA. This HST image revealed it with unprecedented clarity even though it was obtained prior to the First Servicing Mission.

Each of the four gravitationally lensed images is formed by light that took a slightly different path from the quasar to the Earth. While each path is roughly 8 billion light years long, the paths may differ by a light year or so. As a result, when the quasar undergoes a sudden change in brightness, each of the four lensed images will show the same brightness change, but at different times. Precise measurements of these events, and the time lags that separate them, can pin down the distance scale of the gravitational lens system, including both the spiral galaxy and the quasar. That in turn can yield an accurate value for the Hubble Constant, a standard and still uncertain measure of the distance scale of the universe.

Camera: Faint Object Camera
Credit: NASA and ESA

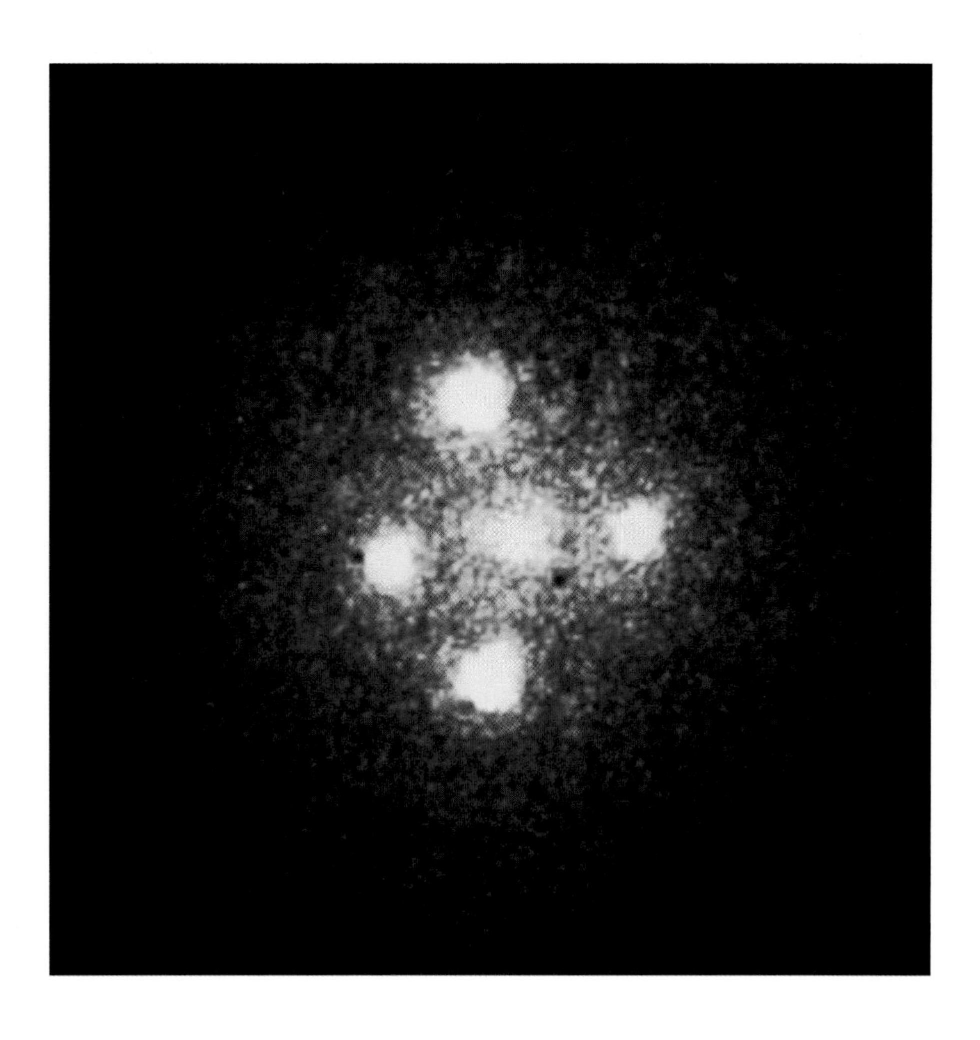

Quasar QSO 1208+101

These are four views of the 18th-magnitude quasar QSO 1208+101, obtained with the HST through filters of various colors. They are shown here in false color, added by means of computer processing. With a red shift of 3.8, QSO 1208+101 was the most distant known object in 1986 when it was discovered with a ground-based telescope. Although large distances in the universe are not accurately calibrated, it must be roughly 10 billion light years from Earth. Several quasars at even greater distances have since been discovered.

As the HST pictures clearly show, there is a second, somewhat fainter light source within one-half of an arc second of QSO 1208+101. The brightness of the second source relative to the quasar is the same when seen through each of the four filters. This strongly suggests that the two sources seen in these images are gravitationally lensed images of the same very distant quasar. If the fainter object were a foreground star, its color would almost certainly be different from the quasar's, and so it would not have the same relative brightness through all four filters. If it were a red star, for example, the fainter object would appear relatively brighter in the image taken through a red filter.

Whatever is acting as the gravitational lens is too dim to be seen in this image and its nature remains unknown.

Camera: WF/PC-1

Credit: J. Bahcall (Institute for Advanced Study, Princeton), and NASA

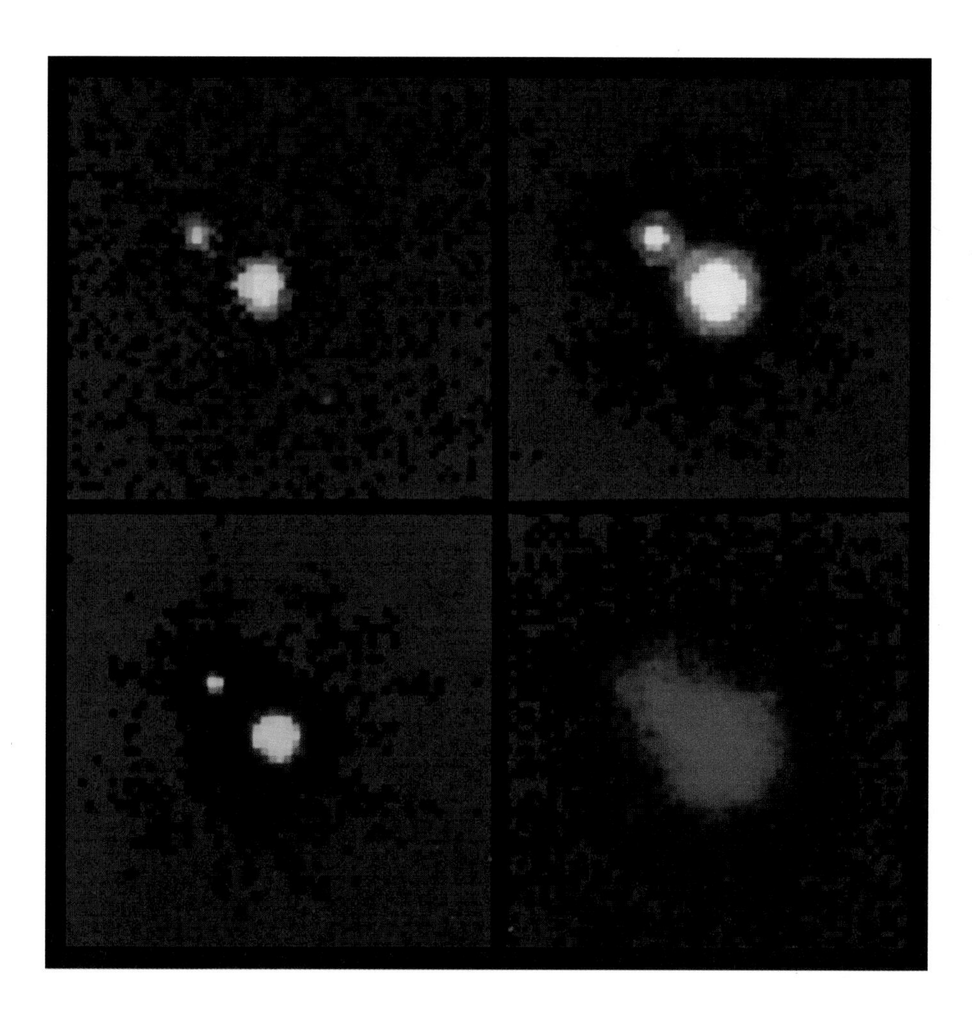

Gravitational Lensing by the Galaxy Cluster AC114

This image shows a distant object which is gravitationally lensed by the cluster of galaxies AC114. It was the result of a six-hour time exposure with the HST. The inset at the upper right shows a ground based photograph of AC114. The rectangle marks the field of view in the HST image.

The lensed object appears to be a bright galaxy located far beyond AC114. The lensing action distorts and splits the light of the distant galaxy so that it appears as the two bright elongated images at upper left and lower right in the HST picture. The two lensed images show a distinct mirror-symmetry and have identical color. The two smaller objects at the center of the HST view are thought to be galaxies in AC114, much closer than the lensed galaxy, and unrelated to it.

In this case, the action of a gravitational lens has magnified the distant galaxy so that it appears larger than it would if it were not gravitationally lensed.

Careful analysis of the HST images such as this can give the distribution of mass necessary to produce the observed gravitationally lensed images. Comparison of that calculated mass distribution with the observed distribution of luminous galaxies reveals how much 'dark matter' must be present in the cluster.

Camera: WF/PC-1

Credit: R. Ellis (University of Cambridge), and NASA

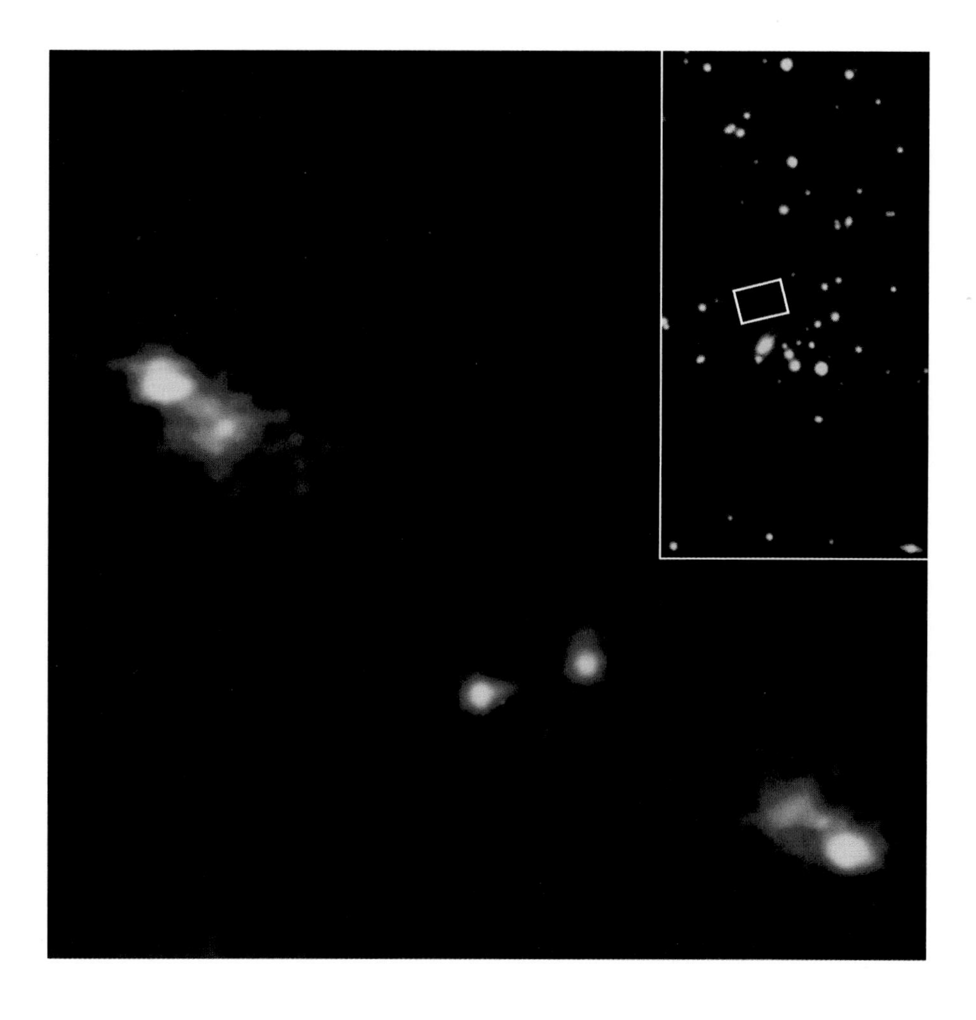

Galaxy Cluster Abell 2218 as a Cosmic 'Zoom Lens'

The pattern of arcs over this HST image of the galaxy cluster Abell 2218 is a spectacular example of the gravitational lensing phenomenon. The arcs are the distorted images of very remote galaxies five to ten times farther away than Abell 2218. These galaxies existed when the universe was only one quarter its present age.

Abell 2218 is such a massive, compact cluster that light rays passing though it are markedly deflected by its enormous gravitational field. The process magnifies, brightens and distorts images of the galaxies far beyond it. In this way, Abell 2218 acts as a powerful 'zoom lens' making it possible to view galaxies so far away that they would not otherwise be visible, even with the most powerful telescopes available. The arcs provide a direct glimpse of how star forming regions are distributed in remote galaxies, and other clues to the early evolution of galaxies. The resolution of the HST reveals numerous arcs that are difficult to detect with ground-based telescopes because they appear so thin. In seven cases, multiple images of the same galaxy can be seen in the HST view.

Camera: WFPC2

Credit: W. Couch (University of New South Wales), R. Ellis (University of Cambridge), and NASA

The Spiral Galaxy M100 in the Virgo Cluster

M100 is one of the brightest spiral galaxies in the huge Virgo Cluster of galaxies, located about 51 million light years from the Earth and estimated to contain 2,500 galaxies in all. M100 actually lies in the constellation Coma Berenices, adjacent to Virgo.

Bright knots of hot young blue stars are seen to delineate the spiral arms of the galaxy. Astronomers who have examined this image in its full detail are excited to note the sprinkling of individually resolved stars and the numerous dark dust lanes across the body of M100. In galaxies so far away, the stars are typically blurred together when viewed by ground-based telescopes. Many of the dark lanes are thinner and more complex than the large dust lanes photographed by telescopes on the ground.

Camera: WFPC2
Technical Information: Composite of three images taken through red, green and blue filters.
Credit: J. Trauger (JPL), and NASA

A Cepheid Variable Star in the Galaxy M100

The high resolution of the HST pinpoints a Cepheid variable star, located in a star-forming region in one of the spiral arms of M100. The top three boxes, from left to right, were taken on May 9, May 4 and May 31 in 1994 and show how the star varies in brightness.

A Cepheid variable is a pulsating star that changes rhythmically in brightness. Times of peak brightness are separated by equal intervals, known as the 'period' of the star. Periods for different Cepheids range in length from hours to days. The brighter the Cepheid, the longer its period. These rare stars act as cosmic 'mileposts', making it possible to deduce the distance to any galaxy in which they can be identified. By discovering Cepheids and measuring their periods, astronomers can determine their true brightnesses. Investigators can then determine the distance of a Cepheid star by comparing its true brightness with its apparent brightness as seen from Earth (or from the HST).

It is difficult to detect individual Cepheids in distant galaxies with ground-based telescopes but, in twelve one-hour exposures over a two-month period, 20 Cepheids in M100 were discovered with the HST.

Camera: WFPC2
Technical Information: Black and white image taken at visible wavelengths.
Credit: W. L. Freedman (Observatories of the Carnegie Institution of Washington), and NASA

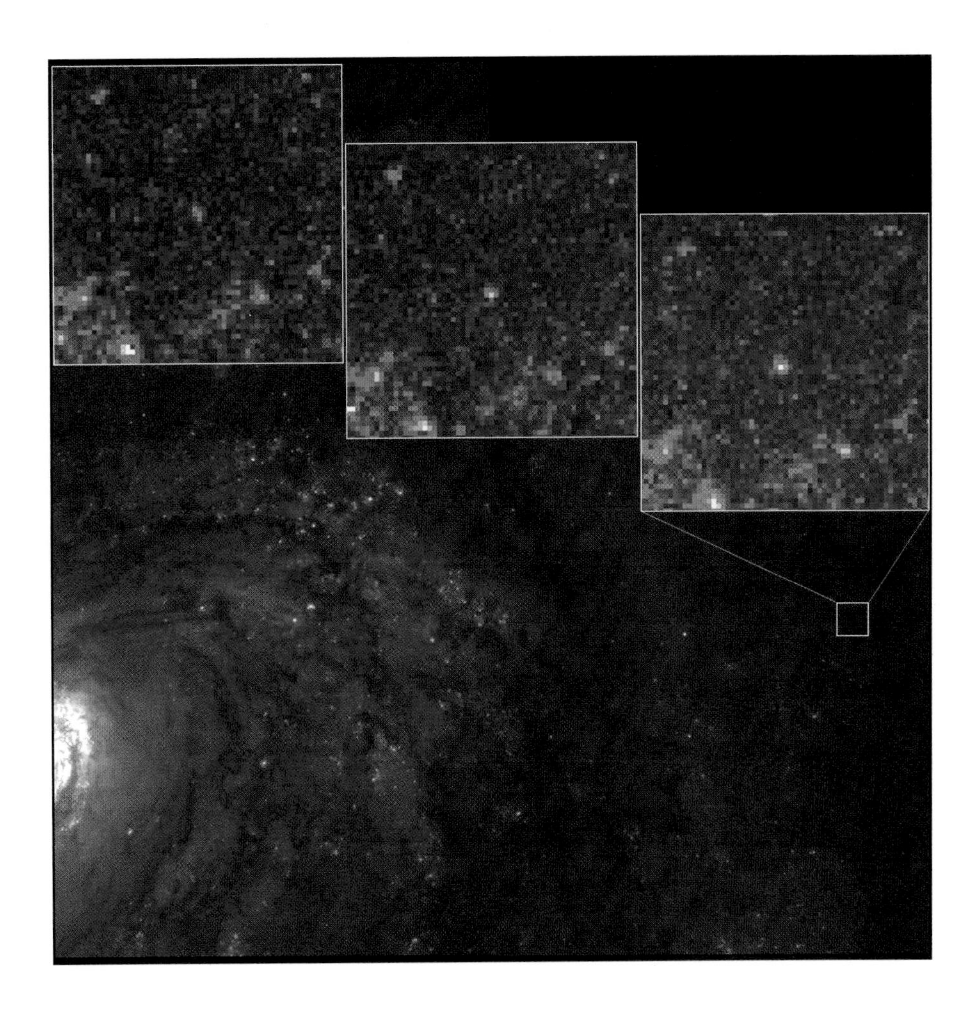

Galaxies of Long Ago

The universe was only two thirds its present age when the light from the galaxies in this image set out on its journey towards Earth. By resolving detail in the images of faint, remote galaxies, HST acts as a 'time machine', allowing astronomers to probe back into the past history of the universe and see galaxies as they were then.

The view here shows the central portion of a cluster of galaxies known as CL 0939+4713. Most of the spiral or disk galaxies visible have odd features, suggesting they were easily distorted by the gravitational influence of neighbors within such a 'rich' cluster. There even seem to be a number of mysterious 'fragments' of galaxies interspersed through the cluster.

Looking back billions of years, we can see that galaxy clusters at that time contained several times more spiral galaxies than similar clusters do now. These spiral galaxies have since disappeared through mergers and disruptions.

Camera: WFPC2

Credit: A. Dressler (Carnegie Institution of Washington), and NASA

Galaxies: Looking Back Through Time

This set of HST images of galaxies shows elliptical galaxies along the top row and spiral galaxies in the rows beneath. From left to right, they are increasingly more distant and so seen as the universe was at progressively earlier times.

The typical elliptical and spiral galaxies on the far left illustrate the two main 'classical' types of galaxy seen in the universe at the present epoch, about 14 billion years since the Big Bang. Both are within a few tens of millions of light years. Elliptical galaxies contain older stars, while star formation continues vigorously in the disks of spirals.

The center left column shows galaxies in a rich cluster when the universe was about two thirds its present age (about 9 billion years old). The elliptical galaxy at the top resembles its present day descendants. By contrast, some spirals have a 'frothier' appearance, with loosely shaped arms.

Distinctive spiral structure appears even more vague and disrupted in the galaxies shown in the center right column. These date from the time when the universe was nearly one third its present age (about 5 billion years old). These spirals are not as symmetrical as the ones we see today and contain irregular lumps where bursts of star formation are taking place. However, even this far back, the shape of the elliptical galaxy in the top box is clearly recognizable.

The galaxies in the far right column existed when the universe was nearly one tenth its present age (2 billion years old). The distinction between spirals and ellipticals is much less clear. Nevertheless, the object in the top frame has all the appearance of a mature elliptical galaxy. This implies that ellipticals formed remarkably early in the universe, while spiral galaxies took much longer to form.

Camera: WFPC2

Credit: A. Dressler (Carnegie Institution of Washington), M. Dickinson (STScI), F. D. Macchetto (ESA and STScI), M. Giavalisco (STScI), and NASA

The Faintest Galaxies Ever Seen

In late December 1995, the HST spent 10 days looking almost continuously at a single small speck of sky in the constellation Ursa Major. The image seen here – called the Hubble Deep Field – was assembled by adding together 342 separate exposures with the WFPC2. The bewildering array of at least 1,500 galaxies includes the faintest ever recorded. At 30th magnitude, these faint objects are about four billion times dimmer than the human eye is capable of seeing. Almost certainly, some of these galaxies are also the most distant ever to be observed.

Astronomers use the term 'deep' for images that record faint objects. The dimmest galaxies include the most distant ones (as well as nearer ones that do not shine brightly). Because their light has taken billions of years to reach the HST, recording the Hubble Deep Field is like using a 'time machine' to peer into the remote past of the universe. Some of the galaxies are being viewed as they were more than ten billion years ago, when in the process of formation.

The area of sky for the Hubble Deep Field was chosen because it is not cluttered with nearby objects, such as stars in our own Galaxy. It is a 'peephole' out of the Galaxy that allows a clear view all the way to the horizon of the universe. Staring at this one spot for ten days, the HST took one picture after another. Each exposure was typically 15 to 40 minutes long. Separate images were taken in ultraviolet, blue, red and infrared light. This enables astronomers to infer, at least statistically, the distances, ages and compositions of the galaxies. Combined, the different colored images give a reasonably 'true color' view.

Camera: WFPC2

Credit: R. Williams and the Hubble Deep Field Team/STScI, and NASA

Acknowledgements

Images and information in this book were acquired with the Hubble Space Telescope by:

Wide Field and Planetary Camera (WF/PC-1) Instrument Definition Team. Principal Investigator: James Westphal (California Institute of Technology).

Wide Field and Planetary Camera 2 (WFPC2) Instrument Definition Team. Principal Investigator: John Trauger (Jet Propulsion Laboratory).

Faint Object Camera (FOC) Instrument Definition Team. Principal Investigator: F. Duccio Macchetto (European Space Agency and Space Telescope Science Institute).

Faint Object Spectrograph (FOS) Instrument Definition Team. Principal Investigator: Richard Harms (Applied Research Corporation).

Goddard High Resolution Spectrograph (GHRS) Instrument Definition Team. Principal Investigator: John C. Brandt (University of Colorado, Boulder).

Corrective Optics Space Telescope Axial Replacement (COSTAR) Team. Project Scientist: Holland Ford (Space Telescope Science Institute).

Special thanks to:

Jennifer Sandoval (Computer Science Corporation) for interpretive art; H. John Wood (Flight Systems and Servicing Project, Goddard Space Flight Center) for data on the HST First Servicing Mission.

Text developed in part from public information materials prepared by Ray Villard and the staff of the Office of Public Outreach, Space Telescope Science Institute.

The Space Telescope Science Institute is operated by AURA (the Association of Universities for Research in Astronomy, Inc.) for NASA, under contract to the Goddard Space Flight Center, Greenbelt, Maryland. The Hubble Space Telescope is a project of international cooperation between NASA and ESA.

The authors

Dr Jacqueline Mitton is based in Cambridge, UK, where she went to study for her astronomy PhD. Since 1987, she has been a writer, editor and broadcaster specializing in astronomy for general audiences. She was appointed the first Public Relations Officer of the Royal Astronomical Society in 1989.

Dr Stephen P. Maran is assistant director of space sciences for information and outreach at the NASA/Goddard Space Flight Center in Greenbelt, Maryland, USA, and the press officer of the American Astronomical Society.